TECHNOLOGY

AN ILLUSTRATED HISTORY OF MACHINES
FROM STONE AXES TO ROBOTICS AND AI

TECHNOLOGY

AN ILLUSTRATED HISTORY OF MACHINES FROM STONE AXES TO ROBOTICS AND AI

Edited by Tom Jackson

Contributors: Tim Cooke · Tim Harris · Tom Jackson · Robert Snedden

Shelter Harbor Press

NEW YORK

Contents

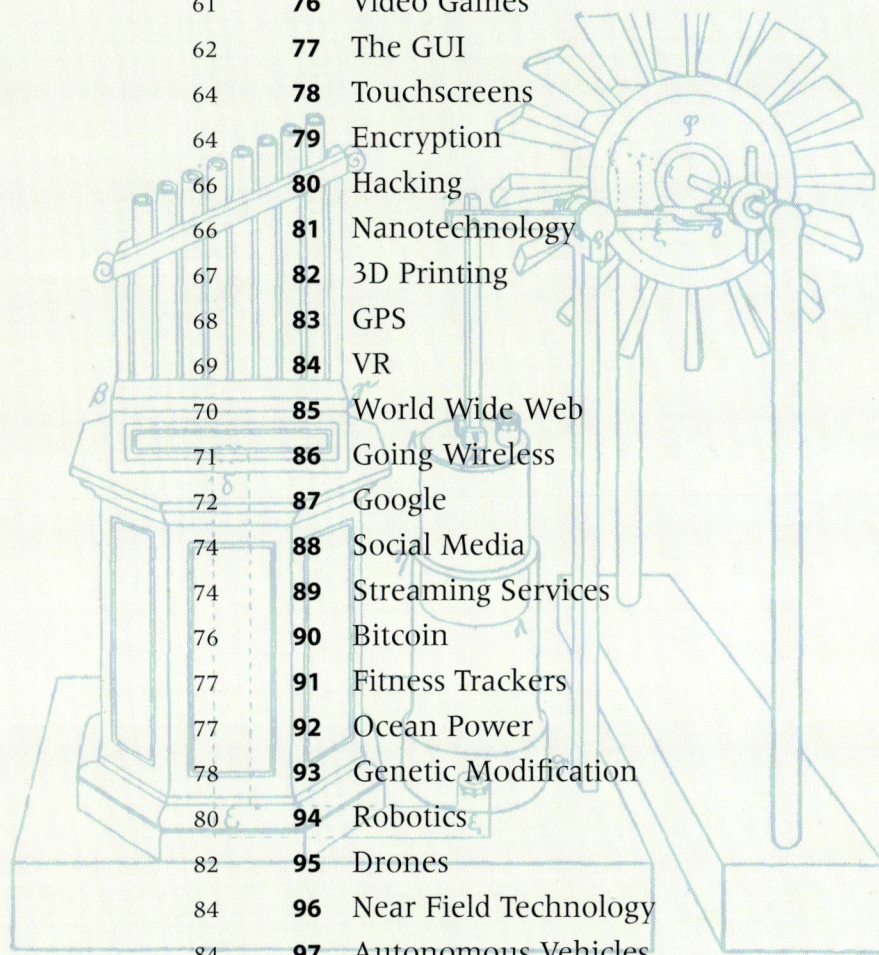

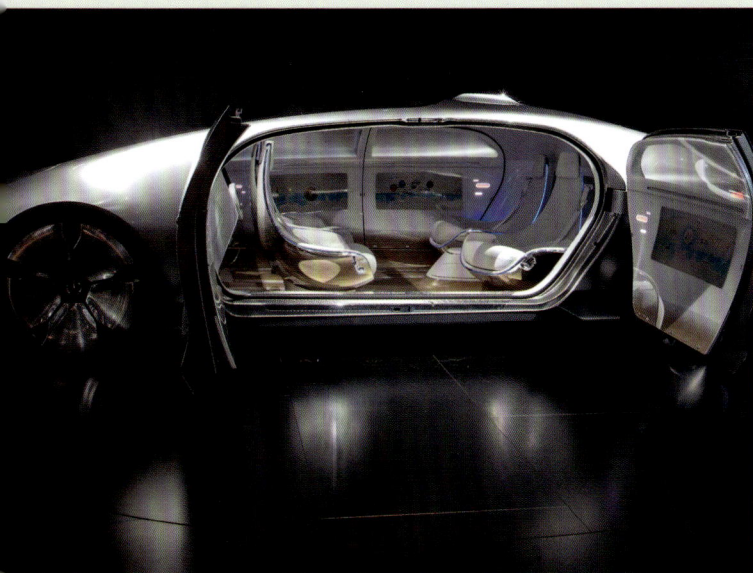

Introduction

IT IS NO EXAGGERATION THAT THE STORY OF TECHNOLOGY IS THE STORY OF CIVILIZATION. WE EVEN DEFINE HISTORICAL ERAS BY THEIR DOMINANT TECH—THE STONE AGE, INDUSTRIAL REVOLUTION, INTERNET AGE—AND THE CONSTANT REFINEMENT IN OUR TECHNOLOGY HAS MEANT WE ARE ABLE TO LIVE LONGER AND HEALTHIER LIVES, WHICH CAN ONLY BE A GOOD THING. AND IT ALL STARTS WITH SINGLE THOUGHT: "CAN WE DO THIS BETTER?"

The thoughts and deeds of great achievers always make interesting stories, and here we have one hundred all together. Each story relates a ponderable, a weighty problem that became an invention and changed our machines, our homes, our minds, and our lives.

The word technology immediately conjures a vision of the latest modern devices that promise to change the future in imaginative ways. Indeed, that is what makes technology so exciting, but it has always been like that. Every generation has invented its way out of the past, and, looking back, we should see all innovations in that light. The word technology roughly translates as the "study of craft," and so a stone ax expertly napped by our prehistoric ancestors is just as much a piece of technology as a silicon microchip being doped in a Taiwanese clean room.

TRACKING KNOWLEDGE

The story of technology is closely aligned with the story of knowledge. As soon as scientists figure out a universal truth about how stuff works, then someone will figure out a way to make use of that new understanding. For example, investigations into strange rocks called lodestones, that attracted and repulsed each other, was the first we knew about magnets. Looking back, this discovery resulted in a cascade of technological innovation. First came compasses, which made it possible to navigate far from home. Then there was the electric motor that used the interplay with electric current and magnets to

The engineers of ancient Rome figured out how simple machines, like wheels and levers, could be used to multiply forces and make work easier.

create motion. If that were not enough, soon after came the electricity generator, which reversed the process of the motor and used motion and magnetism to create electric currents. And this is just one example of many world-changing technologies built on a big scientific discovery. Another was the rhythmic swing of a pendulum, which eventually made it possible to create time technology, which in turn transformed our ability to measure speed and location. Another breakthrough were the invisible electromagnetic waves that shine unseen through the Universe. These waves are best known today by their technological uses and include radio, X rays, microwaves, and infrared.

ACCIDENTAL INNOVATORS

Technological innovations also have a habit of arising by chance as inventors and engineers look for better ways of making and doing. There are many famous examples from modern times, such as superglue, which was intended to be a quick-setting plastic for use in gun sights, but turned out to be much better as an adhesive. Another accidental innovation is Teflon, which was created during research into gases for use in foams and fridges. Instead, the chemists created a super-slippery plastic that now has widespread uses from non-stick kitchenware and self-cleaning fabrics to artificial blood vessels and electrical insulators.

Perhaps the most significant serendipity of this kind comes from the ancient history of metalworking. Few metals exist in a pure form, but the fact that they were locked away in ores would have been a mystery to ancient

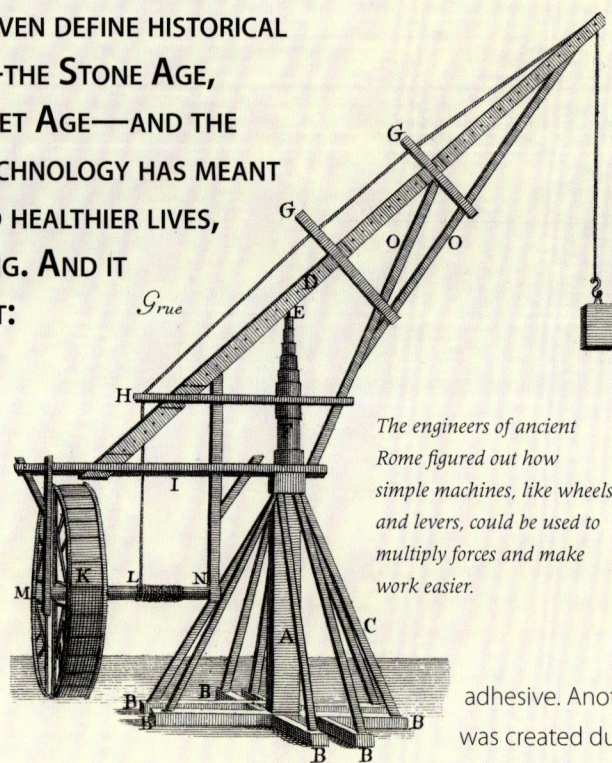

The medieval weapon known as Greek fire, which burned even when wet, was only possible because curious researchers had investigated the properties of natural materials.

craftspeople. Instead, these potters were busy developing hot furnaces for firing clay pots into tough earthenware. After the pots were fired, the craftspeople noticed little blobs of red-hot liquid among the wood ash—or at least this is our best guess of what happened. These droplets cooled down into pieces of hard, shiny metal, which could be bent and flattened, but not easily shattered. What had happened was ore rocks included in the fire by chance had reacted with the burning charcoal and had smelted out metals like copper and lead. Boosting the heat of the fire to release more metals then led to another happy accident. The fiercer fire smelted iron, a much more abundant and useful metal which forms the basis of construction technology to this day.

SEEKING IMPROVEMENTS

Technology is not just driven by chance: there are people who push it along by identifying problems to solve and/or improvements to make. A good example of this is refrigeration technology. Once you've tasted a drink chilled with ice on a hot day, the warm version will never be as satisfactory. Whoever could develop a technology to keep food fresh and drinks chilled would change the world, and today fridges and freezers are in almost every home. The same technology is also used to chill the fuels of space rockets, isolate quantum computers, and preserve living tissue in perpetuity.

The domed ice pits of ancient Persia were able to keep winter ice frozen all through the hot summer. The design used cool winds and flowing water to keep the temperatures inside very low.

Today's visions of the future involve plenty of technology that still needs inventing. Some of that will grow from raw science. It will also develop due to sheer determination to create tech that will meet our future needs. The risks of climate change are driving research into cleaner fuels, like hydrogen, and the means to generate renewable electricity with the power of the wind, sun, and sea.

Another force behind innovation is the desire to make machines that do our work for us. The ultimate technology would be an artificial intelligence that is able to think, reason, and learn how to do whatever we want—or whatever it wants! With each breakthrough, AIs become ever cleverer, outstripping human abilities one by one. The relentless advance in technology begins to sound less appealing. Few innovations are deliberately restricted, but AI innovators themselves are warning that we ought to take steps not to be outsmarted by our own technology. This is not the first time technology has stirred up strong feelings, and you are in the right place to get a bit of perspective. Let's take a look at how smart people have made smart technology through the centuries.

NASA is building AI robots to assist human explorers on spacewalks or in off-world locations like Mars and the far side of the Moon.

TYPES OF TECHNOLOGY

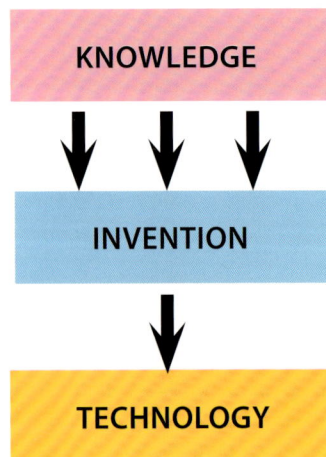

| KNOWLEDGE |
| INVENTION |
| TECHNOLOGY |

"NECESSITY IS THE MOTHER OF INVENTION" IS THE WELL-WORN PHRASE, AND WHILE IT IS TRUE ENOUGH, IT IGNORES THE WAY SCIENTIFIC INVESTIGATION GIVES INVENTORS THE KNOWLEDGE AND MEANS TO CREATE A NEW WORLD-CHANGING TECHNOLOGY.

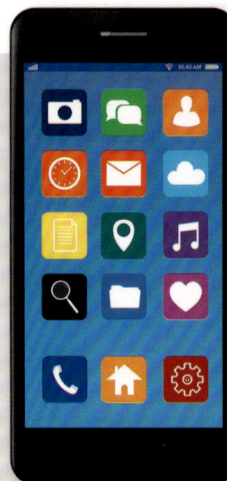

COMMUNICATIONS A fact that is often overlooked is that civilization is based on stories. Our communities are glued together by sharing stories, along with gossip (and, sshhh, secrets, too). A new invention begins like a story, something that is made-up. But, unlike fiction, the tale of a new invention contains within itself the ideas of how to make it real! Our love of stories created an irresistible drive to share ideas, first as images, then writing, and to send this information over long distances. The result is the mail, the telephone, the radio, the TV, the Internet, the Web, social media—oh, yes, and books.

TRANSPORTATION Humans are built to walk (and run in emergencies) and we can only carry what we can hold in our arms (and that includes our children). Obviously, technology that makes moving around and transporting cargo easier has great appeal. While boats were the first means of long-distance travel, ground transportation began with sleds and rollers, and these were then combined into wheeled carts. Next came the transition from muscle power to engine power; eventually engines were made energy-dense enough to drive flying machines. Today's innovation is the drive toward all-electric power.

POWER AND ENERGY Do not underestimate the power of laziness. People have always liked the idea of someone, or something, else doing all the hard work. If other sources of energy could be tapped, our forebears thought, we would become capable of super-human feats. Water and wind were harnessed but remain unreliable. Heat engines, where burning fuel creates motion, arrived in the 1700s. At first these steam engines filled buildings, but now we use tiny electric motors or jet engines the size of a soda bottle.

FOOD AND AGRICULTURE The first human technologies were based on the most primal of needs, the need to eat. For millions of years our ancestors were using stone cutters to slice meat and digging sticks to locate underground water sources. Today's food industry has grown over many thousands of years to tame, emulate, and enhance the natural food chain upon which all other organisms rely. In the last century, farming technology has succeeded in feeding billions of people, but now faces the challenge of doing so without damaging the planet's natural resources.

SPACE In 1957, human technology did something rather incredible. It took us beyond the planet where it had evolved, and made us a space-faring civilization. The know-how for building a rocket and placing a payload into orbit, and then propelling a crew to the Moon, was the sum total of many centuries of enquiry into the nature of space, time, energy, and materials. These were investigated for their own sakes, not as a means to travel in space. Now that we can, we still have much to invent. The human body is not suitable for life in the weightlessness of orbit, and the vast distances of space travel prevent our urge to explore the Solar System. But there are several plans in place—watch this space tech!

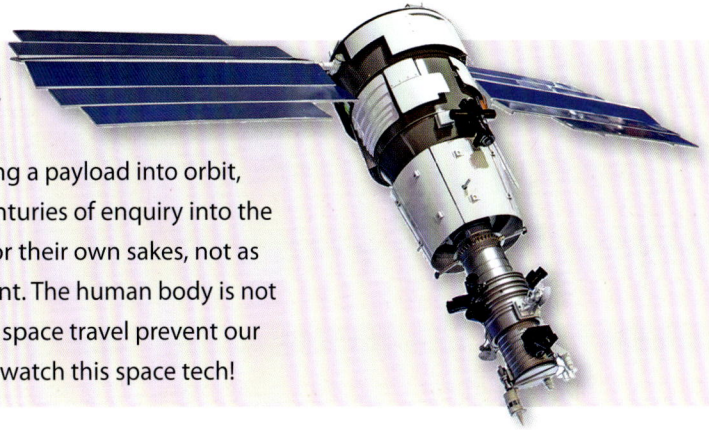

BIOMEDICAL Despite the immense complexity of its systems, a living body is using them to do something rather simple: keep the internal conditions the same and in balance with each other. When this goes wrong, we become sick, and the body needs help. This might be as blunt as building a replacement body part—generally easier said than done, but something we've been trying since the days of ancient Egypt. Next, we take in chemicals that stop us feeling pain or fatigue, and as our understanding of chemicals improves, we create artificial drugs. Soon we may be able to alter our own body systems to end disease—and perhaps even slow the aging process.

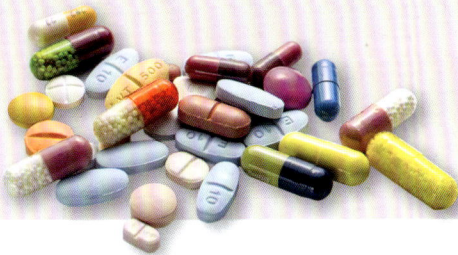

ENTERTAINMENT Life must have been very boring in prehistoric days without video streaming and smartphone games. Or perhaps it was very exciting indeed—as in, utterly terrifying due to the constant threat of a sudden, violent death. Our technology has given us long, safe, and healthy lives, but we crave new things to fill that time. The triple offer of the wandering troubadour—singing, story-telling, and dancing— was superseded by books, cinema, radio, and TV. And today all these modes of entertainment are just "content" delivered on demand to our screens. We'll soon be bored of it, so what's next?

CONSTRUCTION AND MANUFACTURING

One of the defining features of our early pre-human ancestors was their manufacturing technology. Even one million years ago, they were mass-producing stone templates for refining into specific tools when needed later on. From stone and wood we've expanded into metals and plastics, which are stronger, tougher, and lighter. The next big innovation, the division of labor, took many more millennia, and replaced a craftsperson's skill with a team of workers, each performing one task repeatedly. Today, we might get a robot to do that, of course!

ROBOTICS Machines always include a combination of levers, wheels, pulleys, and other simple force multipliers. The field of robotics takes this tech to the next level, so complex machines, perhaps emulating a human form or the body plan of another animal, are able to work independently to perform tasks. This requires a fusion of mechanical and electrical engineering with computer science, so the motion of the robot is able to respond appropriately to changes in the surroundings. To do this, robots are being equipped with sensors and artificial intelligence. Will robots one day be cleverer than us?

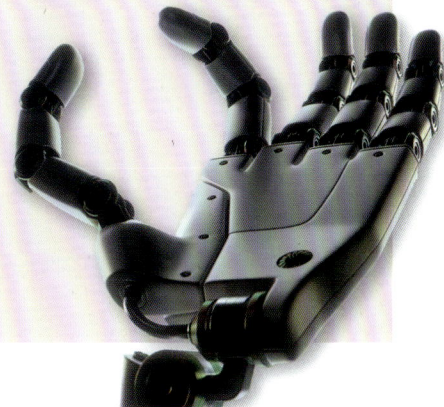

1 Animal Technology

ONCE, IT WAS THOUGHT THAT *HOMO SAPIENS* (MODERN HUMANS) WERE UNIQUE IN THEIR ABILITY TO USE TOOLS, but it is now known that many different animals use and even make their own tools.

Sea otters use rocks to break open the hard shells of abalone and other shellfish. A favorite trick is to lie on their backs in the water, laying the shellfish on their stomachs, and hitting the shells with a pebble.

The most proficient tool-users are chimpanzees, whose range of expertise extends to cracking open tough nuts with stones or chunks of wood, drinking from "sponges" made by dipping clumps of moss in water, and wiping mud from their bodies with damp leaves rather than getting their hands dirty. Chimps in Central Africa use a sophisticated "toolkit" to get at honey, one of their favorite foods. They use a thick stick, or pounder, to break open the entrance to a beehive, a smaller stick to widen different hive compartments, and a spoon-shaped stick to scoop out the honey. Not all chimps' tools are used for peaceful purposes: they have learned the art of warfare, using stones and carefully chosen sticks to throw at rivals.

The bearded vulture hauls bones high into the sky and drops them to smash on rocky outcrops below. It then eats the fragments.

Birds and more

Many birds also use tools. When acorns are plentiful, jays collect and bury them in the ground to feast on later. To remember where it's left its hoard, a jay will sometimes stick a leaf "marker" in the ground next to the acorn. New Caledonian crows select twigs of the right diameter, trim them to the right length, and poke them in tree holes to impale any insect larva living inside. When the crow pulls the twig out, it has a ready meal. Egg-eating Egyptian vultures sometimes find an egg whose shell is too tough to crack. Their solution? Find a stone, pick it up and drop it on the egg until it splits open. They also use sticks to pick up sheep's wool, which they then use to line their nest. Even some ants practice primitive tool-use, picking up tiny stones and dropping them to block the entrance of rival ants' colonies.

A chimp "fishes" for termites with probes made of grass stems.

2 Prehistoric Toolkits

MAKING STONE TOOLS MARKED A MAJOR MILESTONE IN HUMAN EVOLUTION. Previously, hominins had to rely just on the strength and versatility of their hands to collect food and their teeth and jaws to process it.

Anthropologists have found strong evidence in eastern Africa that our early ancestors were using stone tools 3.3 million years ago. At Lomekwi, Kenya, flakes of stone have been matched to stone cores suggesting that they were split off by striking with another stone. The likely toolmakers were *Australopithecus* or *Kenyanthropus platyops* because the age of the tools predates the first species in our genus, *Homo*, by 500,000 years. Similar, but more recent, tools have been found at Gona, Ethiopia, and Olduvai Gorge, Tanzania. The latter, first discovered in the 1930s, gave its name to Oldowan toolkits. Typically, these include hammerstones, which are used to hit stone cores and show signs of battering on their surface; cores, which have flake scars around them; and the sharp stone flakes themselves, which were struck from the cores. Similar tools have been found at other sites in Africa and Asia.

The Acheulean toolkit

Around 1.6 million years ago, a revolution in tool-use took place when *Homo erectus* discovered how to make hand axes. This involved breaking off chunks of rock 5 to 8 inches (12.5 to 20 cm) long, battering them into a pear shape, and sharpening the edges. Tools like this were first found at Saint-Acheul in France in 1859, and the town gave its name to the Acheulean toolkit. Acheulean hand axes were very versatile: they could be used to butcher and skin game, dig in soil, and cut wood. Such a revolutionary technology was bound to spread, and similar hand axes have been found across the world, including in Africa, China, and India.

Although similar simple tools were still being made 250,000 years ago, by this time the pace of tool innovation was accelerating. *Homo sapiens* (modern humans) learned to craft stone points, which could be attached to wooden shafts to make spears for killing larger animals and those that could not be approached closely. Another invention was the awl, a crude needle used to perforate hides. And tools were no longer made only of stone: bone, ivory, and antlers were found to be easier to work with.

An Oldowan cutter (above) and chopper (below). These, the earliest tools, allowed australopithecines and other hominins to cut, pound, and crush meat, bone marrow, and tough fruit and nuts.

The pattern of splits on lithic cores such as these shows that the toolmaker rotated them, indicating that the work was done systematically.

3 Cooking

THE DISCOVERY OF HOW TO COOK FOOD LED TO A REVOLUTIONARY CHANGE IN THE DIET OF OUR EARLY HOMININ ANCESTORS. Cooked food released nutrients, made it easier to digest, and rid plants of any dangerous toxins.

Of course, the essential prerequisite of cooking was fire. Possibly as long ago as 1.5 million years, hominins took advantage of natural fires, "capturing" burning grass ignited by lightning strikes to create heat and light. However, the earliest strong evidence of fire being used to cook is the charred bone fragments found at Wonderwerk Cave in South Africa, which date back 1 million years.

Archaeologists have discovered hearths dating back 790,000 years, with scorched flints and burned seeds and wood, at Gesher Benot Ya'aqov, Israel. The hearths show that their builders could plan as well as control fire. More than 50 different kinds of plants have been found there, including olives, barley, grapes, and nuts, confirming that the local hominin population had a varied diet.

There is plenty of evidence that *Homo erectus* was cooking food 500,000 years ago. Cooking hearths and earth ovens were in general use in Europe and the Middle East 300,000 years ago. By this time, hominins were probably capable of generating their own fire by friction—rubbing hard surfaces until flammable material becomes hot enough to burn—or by percussion, where sparks fly when rocks are struck together. Some 50,000-year-old hand axes from France, made by Neanderthals, were struck by fragments of pyrite, presumably to make sparks.

After the development from hunter-gathering to agriculture 10,000 years ago, cooking became a more sophisticated affair. Tomb paintings from ancient Egypt show how cooking had become something not everyone was responsible for—they depict butchery, baking, brewing, and winemaking as separate crafts.

MATCHES

In 577 CE, Chinese soldiers used wooden sticks coated with sulfur to start fires during a siege. However, the sticks weren't self-igniting. In 10th-century China, the author Tao Gu described people lighting lamps with more refined firelighters made from small pine sticks impregnated with sulfur. Many more centuries passed before, in 1826, English chemist John Walker invented the match, a reliable, self-igniting source of fire with a mix of sulfur and phosphorus on a stick, which was lit by friction alone.

As well as being used for cooking and heat, fires created a focal point for ancient communities.

4 Weapons

EIGHT CAREFULLY CRAFTED WOODEN SPEARS, FOUND IN GERMANY, were made 337,000–300,000 years ago—before the emergence of modern humans, *Homo sapiens*. These spears mark the beginnings of weaponry.

In the hands of a skilled hunter, the Schöningen spears could kill prey at a distance. The later inventions of the atlatl spear-thrower (around 40,000 years ago) and the boomerang increased the effectiveness of hunting weaponry. Although now associated with Australia, the earliest known boomerang was found in a Polish cave, made 23,000 years ago from mammoth tusk. Arrowheads became widespread 20,000 years ago.

By 3000 BCE, people living in Central Asia domesticated horses, revolutionizing hunting and warfare. Horses could carry riders at great speed and over long distances—and pull wheeled vehicles. Horse-drawn chariots were first used around 2000 BCE. In 1274 BCE, 5,000 of them took part in the Battle of Kadesh between Egypt and the Hittites, the biggest chariot battle ever fought. By then, a revolution in metalworking had also taken place, making it possible to manufacture bronze daggers and swords.

The focus for innovation then shifted to China. First the trebuchet, then the invention of gunpowder ramped up the potential for destruction. In 1260, at the Battle of Ain Jalut, Mamluk Egyptian soldiers were probably the first to use hand cannons, the first true firearms. During the Middle Ages, firearms technology became more sophisticated, with the matchlock arquebus used by the Ottoman army around 1440 and muskets in Italy in the following century.

The destructiveness of weapons technology increased dramatically with the industrialization of the 19th century, and this led to wars involving millions of combatants, all well-armed. The development of weapon technology continues today, with the rise of drones that remove human attackers from the combat zone.

TREBUCHETS

For almost 2,000 years, the trebuchet (right) was one of the most formidable weapons of war. Basically a catapult, it used lever technology to hurl rocks or burning projectiles at enemy soldiers or walled defenses during a siege. Traction trebuchets were first used in China in the 4th century BCE. They relied on teams of about a dozen people to pull on ropes to release a rotating arm that could throw a projectile 410 feet (125 m). Counterweight trebuchets, driven by falling weights, were invented in Europe in the 12th century. By the 15th century, they could hit targets 1,375 feet (420 m) away, but, by then, their days were numbered. The era of cannons was dawning.

5 Glue

WHEN EARLY HUMANS LEARNED HOW TO ATTACH FLINT ARROWHEADS TO WOODEN SHAFTS WITH BIRCH BARK TAR, they were doing more than making hunting easier—glues opened up a whole world of possibilities.

Birch bark tar has been called "miracle juice" because it is both incredibly useful and simple to make. The bark is stripped from a tree and, when burned, oil sweats out of it which can be collected on flat stones. Reheated, the oil thickens to a gooey substance, perfect for bonding materials and waterproofing wood. Early humans (*Homo sapiens*) experimented with techniques to develop better bonding agents. The first known compound adhesives, at least 70,000 years old, were found at Sibudu, in South Africa. Their makers used plant gum and red ocher (natural iron oxide) to stick stone blades to wooden ax shafts. Adding ocher to gum creates a less brittle glue when it dries and prevents the gum from disintegrating when wet. Nor were these isolated examples. Careful analysis of hundreds of stone blades crafted by Neanderthals in Italy 55,000–40,000 years ago showed that they had been coated with pine resin glue.

Animal glues

The next big technological breakthrough was the development of animal glues. Boiling skins, tendons, and bones produces collagen, which makes a strong glue. Some 6,000 years ago, Babylonians were using bitumen cement to fasten ivory eyeballs to statues in their temples. Fast forward another two millennia, and the ancient Egyptians used animal-based glues to bond wooden furniture and pharaohs' caskets. There are even stone carvings depicting the preparation and use of glue. The ancient Romans and Greeks made glue from fish skins, bones, and swimbladders, and Chinese artisans boiled fish and stag horns in the 10th century, but in the West glue use declined after the fall of the Roman Empire, only becoming reestablished in the 16th century with the dramatic increase in manufacturing and trade. The Industrial Revolution witnessed rapid advances in glue technology, especially the development of synthetic products to increase their strength, longevity, and water resistance.

ÖTZI THE ICEMAN

In 1991, the well-preserved mummy of a man—later dubbed Ötzi the Iceman—was found in glacier ice near the Italy–Austria border in the Alps. Careful analysis of the remains revealed that he had died between 3400 to 3100 BCE. His belongings revealed much about the clothing, diet, and tools of the time. Among these were a copper ax with a handle made of yew. Birch tar glue had been used to attach the copper blade to the wooden handle. Two arrowheads found with Ötzi were likewise attached to their shafts with glue.

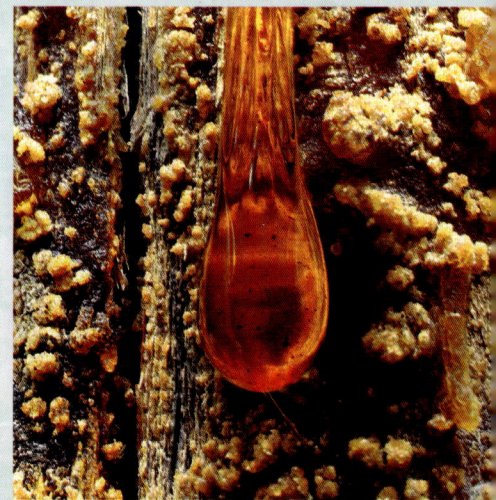

Some trees produce gooey, sticky resins to fill cracks in bark and protect the interior from attack by fungus and insects.

6 Clothing

SCIENTISTS STUDYING THE DNA OF CLOTHING LICE HAVE FOUND THAT THEY BEGAN TO DIVERGE GENETICALLY FROM HUMAN HEAD LICE PERHAPS 170,000 YEARS AGO—strongly suggesting that's when humans began to wear clothes. The problem is that because of the perishable nature of clothing, most—if not all—of the earliest examples have probably been destroyed.

Dyed flax fibers dating from 34,000–36,000 years ago found in a cave in Georgia may have been prepared for use in clothing manufacture. Although it is unlikely that most clothing was preserved for very long, the same is not true for the tools made to craft clothes. A sewing needle made from a long bird bone and dated at 50,000 years old has been found in Denisova Cave, Siberia, and there are more recent examples from China, France, Slovenia, and elsewhere in Russia. The discovery in a French cave of an ivory statuette of a partially naked woman (the Venus of Lespugue) was particularly significant: she is wearing what appears to be a twisted fiber skirt. Dated at 24,000–26,000 years old, this is the earliest depiction of someone wearing clothes.

Clothing created a new habitat for blood-sucking lice to set up home on the human body.

Weaving cloth

Simple warp-weighted looms have been found at sites used by people of the Starcevo Culture in Southeast Europe between 6200–4500 BCE. In such looms, the warp threads hung from a wooden bar and were held taut by stone loom weights. Their use shows that textile manufacture had become an important part of the life of communities at this time. Similar looms were later employed in ancient Greece and subsequently across Europe. The rise of Middle Eastern and Mediterranean civilizations witnessed the development of more sophisticated means of making clothing. Spinning and weaving linen from flax has a long history in ancient Egypt. Both this and wool manufacture from sheep and goats were features of society in Mesopotamia, Egypt, Greece, and Rome by the first millennium BCE.

The Venus of Lespugue, a clay figure from at least 24,000 years ago, appears to be wearing clothing.

SHOES

Since they would have been made from highly perishable materials, the earliest footwear has probably long since vanished. The oldest shoes known are dozens of carefully crafted sandals, made from sagebrush bark, that were found under a layer of volcanic ash in Fort Rock Cave, Oregon. Most date from 8500–7200 BCE. They have closed toes and tough, twined soles, with weft fibers twisted around the warps. The oldest leather shoe (right) is a simple affair, a single piece of cowhide laced with leather cord along the front and back. Discovered at the Areni caves, Armenia, it dates from 3500 BCE. Ötzi the Iceman, whose well-preserved possessions dating from 3400–3100 BCE were found in the Alps, had sophisticated shoes, with bearskin soles, deerskin side panels, and a bark-string net that pulled tight around the foot. Thong sandals of woven papyrus or palm were worn in ancient Egypt before 3000 BCE.

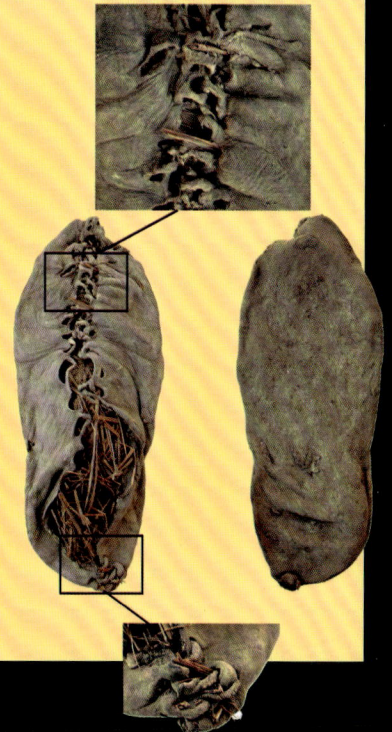

7 Numbers

HUMAN CIVILIZATION COULD NOT HAVE ADVANCED WITHOUT NUMBERS. When early hominins first used their fingers to count the number of animals they had killed, or how many children they had, they were taking the first baby steps on the road to being able to describe the world numerically.

THE ABACUS

The first abaci, or counting frames, were used in Mesopotamia between 2700 to 2300 BCE, and were found in ancient Egypt, Persia, Greece, China, and Rome in the following centuries. The first "pocket calculator," an abacus small enough to fit in a pocket, was invented in Rome in the 1st century CE. This device enabled calculations up to and including millions. It consisted of grooves cut into a wooden or brass plate. Seven long grooves held up to four beads (representing 1, 2, 3, 4; 10, 20, 30, 40; 100, 200, 300, 400; etc) and seven short grooves held one bead (representing 5, 50, 500, etc). An additional groove may have counted halves, thirds, and quarters.

Millennia-old tally sticks found in Africa are the first indicators that people were making simple tools to help them count. The Ishango Bone, discovered in D.R. Congo and 20,000–22,000 years old, is a baboon's leg bone with three columns of short lines cut into it. Although its exact significance is not known, each incision presumably represents a unit. Even older, at 44,000 years, is the Lebombo Bone from South Africa, which has 29 notches carved into it. Some anthropologists believe that the incisions represent the number of days in a lunar month, but this is highly speculative.

The Compendious Book on Calculation by Completion and Balancing, written by Muhammad ibn Musa al-Khwarizmi around 820 CE, was a landmark in the study of math, including algebra.

Using symbols

The advent of written systems marked the next great revolution in the story of numbers. The ancient Egyptians invented a base-10 system some time before 2700 BCE. It has seven hieroglyph symbols for the numbers 1 (a single stroke), 10 (cattle hobble), 100 (coil of rope), 1,000 (lotus plant), and so on, and could also express fractions. There was no concept of place-value, however, so while a single hieroglyph character represented 1,000,000, 54 characters were required for the number 999,999!

An alternative way of counting, which developed in Babylon, was a base-60 system whose legacy persists in the division of a minute into 60 seconds and a circle into 360° (6 x 60). The Babylonians' other great invention was algebra. The Plimpton 322 clay tablet, which was inscribed in 1800 BCE, has solutions to "Pythagorean triples," the integers that solve the equations $a^2 + b^2 = c^2$. From the 1st century BCE, the Romans used combinations of seven letters from their alphabet (I representing 1; V for 5; X for 10; L for 50; C for 100; D for 500; and M for 1,000) for their numeric system. Like the earlier Egyptian system, it was

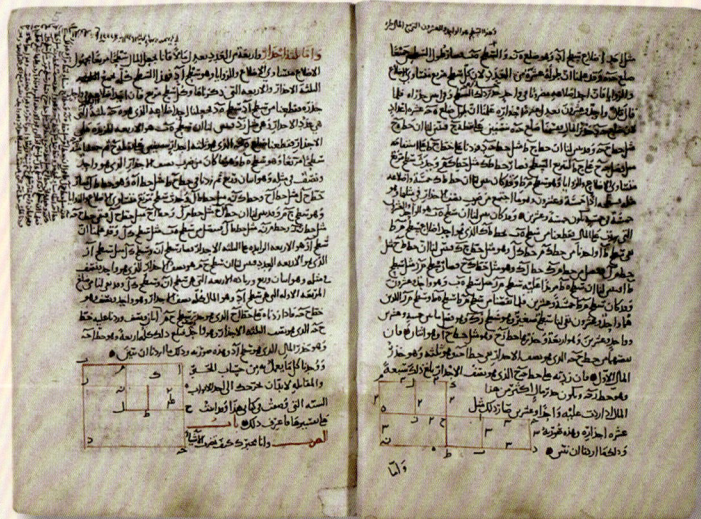

The Ishango Bone is probably more than just a record of simple numbers. It may be a device that aids calculations.

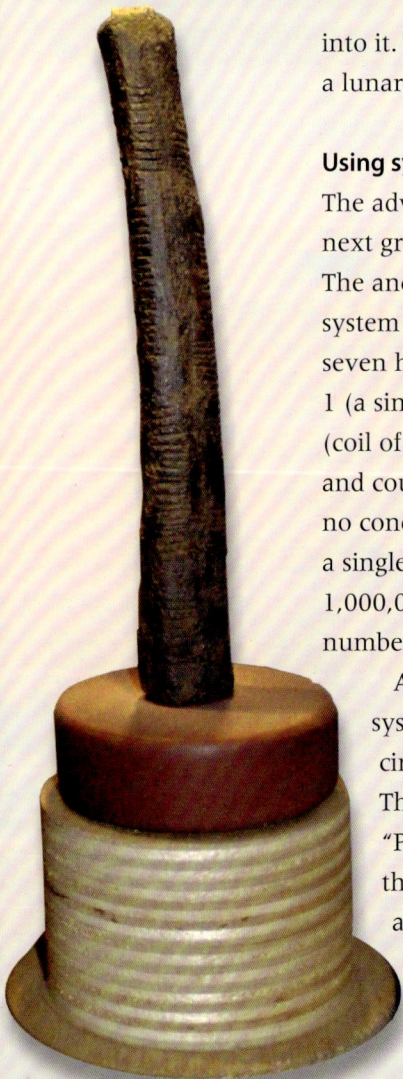

ZERO

A zero-like symbol was used by the Babylonians in the 3rd century BCE as a placeholder, and by the Romans to indicate a lack of quantity. However, it was the great 7th-century Indian mathematician Brahmagupta who first explained that 0 was actually a number. It may seem obvious now, but he was the first to show that deducting a number from itself produces 0, and that the product of any number with 0 produces 0. The use of 0 as a digit in its own right spread to China, then Persia and the Middle East, finally reaching Europe in the 12th century.

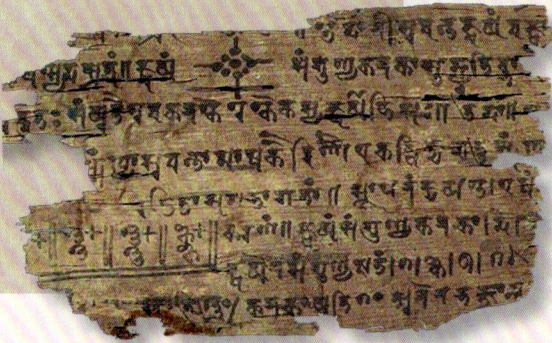

extremely unwieldy, but it remained dominant in Europe until at least the 12th century.

The familiar Hindu–Arabic numbers long used in most parts of the world today were invented in India in the 6th or 7th century. This base-10 system is easier to use than its earlier counterparts because it incorporates place-value—there are separate columns for ones, tens, hundreds, and so on, making complex calculations relatively easy. This system gradually spread west, reaching Europe by the 12th century through the writings of mathematicians such as the Persian al-Khwarizmi, and the Italian merchant Leonardo of Pisa, better known as Fibonacci.

8 Rope and String

BY TWISTING LENGTHS OF PLANT FIBER AROUND EACH OTHER, people found they could make strong rope for pulling, lifting, and attaching heavy items. The same principle enabled them to make a variety of bow strings for hunting and bow drills for making fire.

Rope has a very long history, as shown by the discovery of a 40,000-year-old ivory rope-making tool in southwest Germany. The device is 7.9 inches (20 cm) long, carved from mammoth ivory, and has four holes about 0.3 inches (8 mm) in diameter bored in it, each lined with precisely cut spiral incisions. Researchers believe people threaded plant fibers through the holes to twist and strengthen them into rope. Impressions of rope have also been found on fired clay that is 28,000 years old, and fossilized two-ply rope 17,000 years old was found in one of the Lascaux caves in France.

String was used for bow drills, important for making fire, in at least the 5th or 4th millennium BCE by people of the Indus Valley civilization and in Iran. String would also have been used for fishing lines, bow strings, nets, and traps. In ancient Egypt, between 4000 and 3500 BCE, reed and palm fibers, flax, and papyrus were used to make rope, without which the construction of the pyramids would not have been possible. Hemp fiber was used in China as long ago as 2800 BCE.

A rope from the Mary Rose, *an English warship that sank in 1545 and was raised from the seabed in 1982.*

9 Pottery

AROUND 9000 BCE, AS THE FIRST FARMING COMMUNITIES FORMED, the need for containers grew—especially to carry water. Although pottery had been made before this time, it now became a regular feature of many societies.

The Venus of Dolní Věstonice is a ceramic statuette of a female figure dated between 29,000 and 25,000 BCE. Discovered near Brno in the Czech Republic, it was made from clay fired at a relatively low temperature. The earliest known pottery vessels were made in China: fragments have been found at Xianrendong, dating from about 18,000 BCE, and Yuchanyan Cave, about 2,000 years more recent. At this time, clay would have been kneaded to remove the air, then shaped, dried, and fired. The firing process involved placing the clay items in a hollow in the ground—a makeshift kiln—then lighting a fire on top. By 9000 BCE, clay-based ceramics were being produced to make tiles and bricks, as well as containers for water and food, and the technology spread out from East Asia. Around 3500 BCE, the slow wheel was introduced, allowing potters to rotate the clay to make round pots.

Higher temperatures

From about 2600 BCE, advances in kiln technology in the Indus Valley led to hotter temperatures and better quality, non-porous stoneware ceramics. Another quality revolution occurred in China about 600 CE when white kaolin clay was mixed with a type of granite and fired at temperatures above 1,200°C (2190°F) to produce lightweight porcelain. Such high-quality ceramics were produced in large quantities during the Han dynasty (618–907), both for decorative and functional purposes. Like stoneware before it, porcelain production spread out from East Asia to other parts. However, unlike other forms of pottery, bone china has its origins in the West. In 1748, Thomas Frye started manufacturing it from clay, granite, and animal bone. It is a very tough, translucent form of porcelain. Until the second half of the 20th century it was only manufactured in the UK.

The combination of the kaolin and extreme heat causes the clay mix in porcelain to become a vitrified (glasslike) mineral called mullite. Because of mullite's high melting point (1,840°C; 3344°F), porcelain makes good heat-resistant containers. It also glazes and paints well.

10 Bread

AS ONE OF THE FIRST FOODS HUMANS MADE—AS OPPOSED TO FOOD THAT WAS **KILLED OR COLLECTED**—bread has been a dietary staple in many parts of the world for millennia. It is relatively easy to make and carry.

Thousands of years before the dawn of agriculture, people used wild plants to make flour. About 14,400 years ago, still long before grains were cultivated, Natufian hunter-gatherers were baking flatbreads on hot flat rocks in Jordan. Archaeologists found charred mixed-grain breadcrumbs in two stone fireplaces. Lab analysis showed that the bread had been made from wild oat, barley, and einkorn wheat flour. It is likely that the fireplaces were dual-purpose, used to roast meat over the flames and bake dough in the hot embers.

Unleavened breads are still the norm across North Africa as well as Western and South Asia.

People discovered that fine flour made better bread, and by 800 BCE Mesopotamians were milling, or grinding, grain between massive, circular, flat stones to produce powder-fine flour.

Cereal revolution

Bread-baking really took off in the Fertile Crescent of the Middle East, where wild cereals were cultivated from 9000 BCE. Over the course of the next four millennia, the cultivation of plants began in North America, the Indus Valley, and China. Bread became an important part of people's diet in many parts of the world. Ancient Egyptian hieroglyphs from 3000 BCE show dough rising in ovens.

The next technological breakthrough was the discovery of leavening to make a lighter and fluffier bake. This may have been discovered by accident, perhaps when fungal yeast grew on dough left unprotected on a hot day. The dough would have started fermenting before it was baked, producing superior bread. The use of a sourdough "starter" became the norm and by 300 BCE, the skilled bakers of the Nile Valley were practicing commercial yeast production. The professional bakers of ancient Greece and Rome produced a wide range of breads and pastries baked in enclosed ovens with flues.

11 Agriculture

THE NEOLITHIC, OR AGRICULTURAL, REVOLUTION STARTED AROUND 12,000 YEARS AGO as Earth began to warm up after the last Ice Age. It would forever change the way people produced food, and paved the way for civilization.

PLOWS

Early farmers understood the need to loosen soil in preparation for sowing seed or planting crops. They used wooden digging sticks or hoes, but this was a very inefficient way of preparing large fields. When oxen were domesticated in Mesopotamia and the Indus Valley in the 6th millennium BCE, they were used to pull simple wooden plows called ards. The animals dragged the ards, which traced shallow furrows. The oldest known furrows were unearthed by archaeologists in the Czech Republic and date to 3500 BCE. Around 2300 BCE, the Assyrian and Egyptian civilizations invented iron ards, which cut deeper traces in the ground. Two millennia later, Han-dynasty farmers in China used iron moldboard plows, which could turn even the heaviest clay soils.

At that time, people began to cultivate wheat, barley, peas, lentils, and flax in the Fertile Crescent of the Middle East. They became farmers. Not long after, arable farming began in the Indus Valley, bananas were cultivated in New Guinea by 7000 BCE, rice was grown in China by 5700 BCE, and potatoes cultivated in Peru by 3600 BCE.

New techniques were invented to make farming more efficient. Newly domesticated oxen were used to plow the land, and this technology dramatically increased the area that could be farmed. Irrigation—known from the 6th millennium BCE in Khuzestan, Iran, opened up lands previously too arid for crops. Around then, crop rotation—alternating the planting of cereals and legumes—was practiced in Mesopotamia; although the chemistry of this couldn't have been understood, experience had shown that it led to higher yields of both crops. The use of animal manure as fertilizer was supplemented by wood ash in many areas and bird droppings, or guano, in South America.

In the agricultural heartlands of southwest Asia, North Africa, and southern Europe, cereal crops were cut with sickles, dried in shacks, and threshed by trampling animals; the grain was stored in granaries and used to make bread and beer. Olives were grown for olive oil and flax for the manufacture of linen cloth. Peas, lentils, figs, dates, apples, and pistachios were grown where local conditions suited their cultivation. Farming productivity increased over time. In the Roman Empire toward the end of the first millennium BCE, it was probably five or six times higher than it had been in the pre-agricultural epoch. That meant only a small fraction of society was needed to produce its food requirements—freeing the rest to play other roles.

The ancient Egyptian culture was built around farming the fertile land beside the Nile River.

12 Livestock

THERE WERE TWO ASPECTS TO THE AGRICULTURAL REVOLUTION, WHICH BEGAN ABOUT 12,000 YEARS AGO in the Middle East: people began planting crops and living a sedentary lifestyle, and they began to domesticate animals.

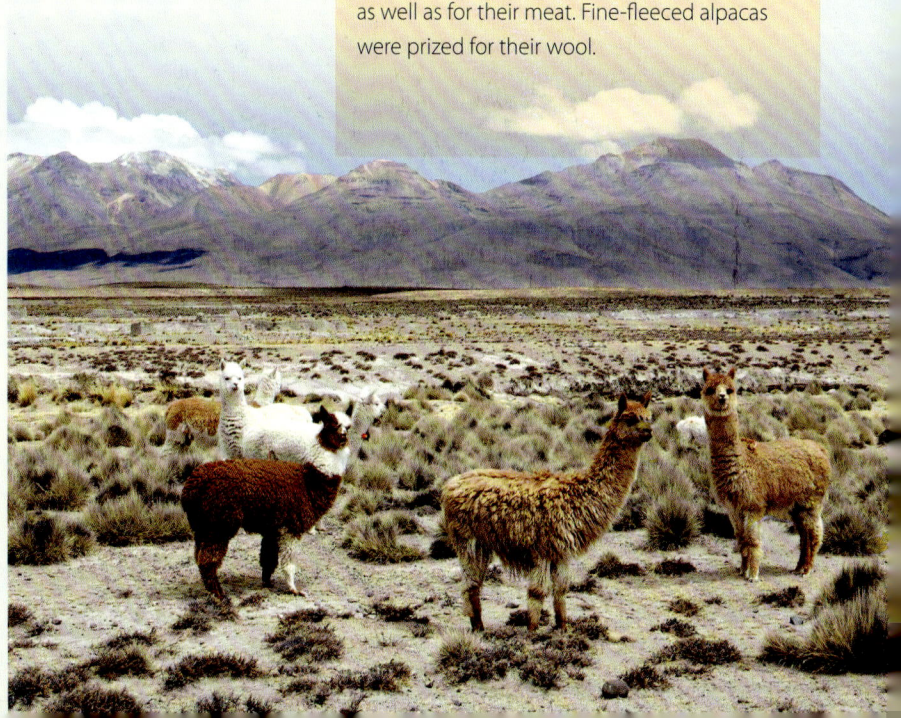

Wild animals, such as this Babylonian goat, weren't domesticated overnight. Generations of selective breeding were needed to produce the docile pigs, sheep, and goats that we are familiar with today. But once in captivity they became an invaluable resource for meat, wool, and milk.

As people began to cultivate crops and establish settled communities, there was less time to hunt. Hunting was a hit-and-miss activity; it could be hard to make a kill sometimes, while on other occasions the meat from a successful hunt was wasted because there were not enough mouths to eat it. For a settled community, it made sense to breed captured animals so they could be slaughtered on demand.

Wild boar were kept captive from at least 8500 BCE in the Middle East and not long after in ancient China. Wild sheep and goats became the livestock of choice for nomadic pastoralists, people who hadn't yet settled down but were moving away from a hunter-gatherer lifestyle.

Beasts of burden

The most significant breakthrough was the domestication of wild cattle in the Middle East, South Asia, and Southeast Asia from the 6th millennium BCE. As well as providing meat, milk, leather, and manure to fertilize crops, oxen were strong draft animals, capable of transporting loads and pulling plows—which greatly extended the area that could be cultivated. When wild horses were first captured and bred on the steppes of Central Asia in around 3000 BCE, it was probably to secure a source of meat, but it must soon have been obvious that they were a superb form of transport. They could carry people and provisions over long distances and at great speed over short distances. Horses later became an important factor in warfare, able to carry warriors and pull chariots. By the end of the 1st millennium BCE, asses, camels, poultry, elephants, rabbits, and guinea pigs had been domesticated in various parts of the world.

ALPACAS

From about 5000 BCE, people in the Andes Mountains began to switch from hunting guanacos and vicunas to herding. They captured the animals and bred them. Over time, selective breeding produced llamas from guanacos and alpacas (below) from vicunas. Evidence from Telarmachay in Peru shows that by 1800 BCE these animals made up three-quarters of people's diet. The larger llamas were bred as pack animals, a role they still play today, as well as for their meat. Fine-fleeced alpacas were prized for their wool.

13 Brewing

ALTHOUGH BREWING BEER IS NO SIMPLE TASK, IT HAS A VERY LONG HISTORY —possibly even predating the agricultural revolution and the earliest settled communities.

A master brewer oversees beer production in Babylon.

Reliefs on tombs dating from 2400 BCE provide a good insight into how ancient Egyptian brewers went about making beer. They crushed partly germinated barley, mixed it with water, and then dried it to form cakes. These were then broken up and mixed with water to produce an extract that was fermented by wild yeasts naturally present inside pottery vessels. Yet there is evidence that people may have been brewing long before the Egyptians. Chemical analysis of stone mortars found at Raqefet Cave in Israel showed signs of grain-based alcohol fermented around 11,000 BCE by the Natufians, a group of hunter-gatherers who had a sedentary lifestyle.

A beer known as *kui* was brewed in northeast China in 3000 BCE. Analysis of pottery containers there has shown various ingredients, including barley (which may have been introduced into East Asia especially for brewing), millet, yam, and the roots of lilies that were used in the process. And traces of wine and a barley-based beer were found in ceramics dated between 3100 and 2900 BCE in the Sumerian settlement of Godin Tepe. Beer was an essential part of daily life in Egypt, drunk in great quantities at religious festivals and given to laborers working on the pyramids. Later, the ancient Greeks and Romans favored wine over beer, considering the latter a barbarian drink.

14 Metals

HUMANS HAVE KNOWN ABOUT SOME METALS FOR MILLENNIA BECAUSE THEY ARE FOUND PURE IN ROCKS. Being soft, gold and silver were useful for decorative purposes, but copper was a different story.

A lead coin from medieval India. Lead was one of the first metals to be smelted from ore. Pure lead is too soft to be of much use, although it was used for piping and waterproofing until recent decades.

Copper knives and sickles are known from 7000 BCE in Anatolia after people realized that pure copper found in rocks became harder if hammered and could be shaped. No one knows how people first discovered that metals could be created by heating certain rocks. The knowledge probably came from lucky accidents—maybe a fire lit on a chunk of ore. Once the discovery was made, smelting became a crucial technological breakthrough. Archaeologists in Serbia have found evidence

that copper ore was smelted at high temperatures to create metallic copper as long ago as 5000 BCE. Surely another fortuitous accident was the discovery that molten copper and tin combine to form bronze, an alloy stronger—and sharper—than both constituents. This revelation kick-started the Bronze Age around 3300 BCE.

Iron smelting occurred later in the 3rd millennium BCE, since it required hotter furnaces. The discovery of steel may have been another happy accident—fragments from a Hittite site in Anatolia dating to 2000 BCE show traces of carbon, possibly from charcoal in a smelting furnace. The addition of carbon made the resultant metal stronger—and tougher than bronze. In around 1200 BCE, iron manufacture increased dramatically, possibly because a shortage of tin made bronze production difficult. This marked the start of the Iron Age. Iron was ideal for making sharp implements, especially weapons.

Çatalhöyük in southern Turkey was one of the first cities in history. It was built about 9,000 years ago by a civilization that was transitioning from stone tools to copper ones.

15 Irrigation

THE FIRST GREAT CENTERS OF AGRICULTURE WERE THE RIVER VALLEYS OF THE MIDDLE EAST, where sometimes there was too much water and sometimes too little. The answer was water management, or irrigation.

In the 6th millennium BCE, gangs of Samarran people dug ditches up to 4.8 km (3 miles) long in the arid country around Choga Mami, in present-day Iraq, to channel water from a nearby river so they could grow wheat and barley. Farther east, in the Indus Valley, people constructed reservoirs to store floodwater to use in the dry season. Around 3000 BCE, farmers in Mesopotamia invented the *shadoof*, a bucket on a rope hanging from a pivoted wooden pole. They used this device to transfer water from a canal onto dry soil. The technology was later used in ancient Egypt, where they developed the basin irrigation technique. Fields were divided by a criss-cross network of earthen banks, or dikes. When the river flooded, water was allowed to fill some basins to saturate the soil beneath and deposit its nutrient-rich sediment. The water would then be allowed to flow into the next groups of basins.

Ancient Egyptians used shadoofs to raise river water and took advantage of the Nile's annual flood between June and September.

16 The Wheel

THE ROLLING WHEEL DIDN'T APPEAR UNTIL LONG AFTER MANY OTHER TECHNOLOGIES. The tricky thing wasn't conceiving of a cylinder rolling on its edge, but working out a way to connect this to a stable platform.

ROLLERS

Have you ever wondered how ancient Egyptian builders moved the giant blocks of stone to build the pyramids? Or how the huge stones of Stonehenge were put in place in 3000 BCE? One theory is that the stones may have been pulled on wooden rollers. By placing tree trunks of roughly similar diameter side by side, laying the blocks on top of them, and using a combination of ropes and levers to pull and push, they could have been moved many miles. Not all archaeologists are convinced by this theory, however.

The solution was a wheel-and-axle combination, but this wasn't viable until metal chisels—initially of copper—were invented to craft the precise fittings needed. Before the invention of the rolling wheel, potters used simpler wheels to rotate their clay as they worked it, thereby producing rounder, smoother pots.

The first wooden cart wheels were probably made from a single piece of wood, a section of a tree trunk. More advanced wheels were crafted from several attached and rounded planks such as one found near Ljubliana, Slovenia, dated from at least 3150 BCE. It was constructed from two planks of wood held together with four cross braces and was attached to a wooden axle, which would have rotated with the wheels.

Wheels of war

While speed and maneuverability were not important for wooden carts pulled by oxen or pushed by men, these qualities were vital for vehicles of war. By 2000 BCE, warriors of the Sintashta culture on the steppes of Russia were fighting in lightweight chariots pulled by two horses. Within 700 years, these warcraft were known in Egypt, Mesopotamia, and other parts of the Middle East. Chariot wheels were later equipped with robust bronze or iron tires.

The critical innovation that made high-speed, horse-drawn war chariots viable was the invention of the spoked wheel, which is strong but lightweight.

17 Sailing

PEOPLE HAVE USED SIMPLE BOATS PROPELLED BY CURRENTS OR OARS FOR MILLENNIA. The invention of the sail made it possible to exploit the immense power of wind.

A 7,000-year-old painting from Mesopotamia is the earliest evidence of a sail. It shows one on a reed-bundle boat afloat on the Persian Gulf. Another painting, this time from Egypt and dating to around 3500 BCE shows a similar boat on the Nile. Between 3000 and 1500 BCE, Austronesians built the first truly ocean-going sailing vessels to colonize large parts of the Indo-Pacific. From the 5th century BCE, Phoenician, ancient Greek, and Roman navies fought for domination of the eastern Mediterranean with triremes, war craft with banks of oarsmen and two large square sails. Square sails couldn't use crosswinds, so human labor regularly had to take over. In the 15th century, the Portuguese built fast caravels, wooden ships with several lateens (see box), for great voyages of discovery.

The "age of sail" peaked in the 19th century when fast ships had up to 32 sails with an area of 2,976 square meters (32,000 square feet) giving a maximum speed of 32 kmph (20 mph).

THE LATEEN SAIL

The invention of the triangular lateen sail was one of the biggest leaps in sailing technology. This enabled boats to "tack" into the wind to take advantage of crosswinds. The first lateen sailboats are thought to have been introduced to the Mediterranean by sailors from the Persian Gulf in the 2nd century BCE. The lateen sail was rigged along the length of a boat, and could be swung left or right to take advantage of the prevailing wind. This gave the vessel forward movement when other sailboats were becalmed. By the 6th century CE, the lateen sail had replaced the square sail throughout the Mediterranean.

The remains of an ancient Nile riverboat in a Cairo museum. Boats float downstream easily but the sail is needed to go upstream, and northerly winds on the Nile boosted commerce and construction in ancient Egypt.

18 Writing

THE ORIGINS OF WRITTEN COMMUNICATION CAN BE SEEN IN ROCK ART HUNDREDS OF THOUSANDS OF YEARS OLD. These were pictograms that depicted things to convey simple messages, but they could only communicate basic ideas. Writing is believed to have developed independently in Mesopotamia, China, and Mesoamerica.

The written language invented by the Sumerians of southern Mesopotamia enabled communication over long distances and through time. Their writing developed from a counting system of shaped clay tokens, which represented different goods. The first commodity transactions were recorded on unmarked tokens but, by 3350 BCE, people were inscribing simple symbols (pictograms) onto wet clay with styluses made from cut reeds—cuneiform writing. The symbols represented items, and numbers, such as quantities of grain, beer, or livestock, could be recorded. Sun-dried tablets could later be soaked in water for reuse. Tablets with more important data were fired in kilns.

Pictograms were fine for describing financial transactions or lists of things, but were not much use for communicating more complex ideas. For example, an image of a goat with four strokes across it and a temple suggested that there were four goats in a temple, but it didn't explain whether the animals lived there or were traded there, whether they were dead or alive, or who owned them.

Texts dated to 3200 BCE found in the Sumerian city of Uruk have a mix of pictograms and phonograms, elements that represent sounds. Sumerian scribes gave phonetic values to signs based on how a word sounded. If a mouth symbol represented the word "ka," that symbol would be used wherever a word included the syllable "ka"—even if the word

IDEOGRAMS

Ideograms represent ideas through symbols and are independent of any language. They developed from pictograms, which depict objects and whose origins stretch back to Neanderthals' depictions of animals hundreds of thousands of years ago. By 15,000 BCE, the cave artists of Lascaux, France, had taken things a step further and were using ideograms. Their wall paintings of speared animals convey the message that hunts had been successful. Today, ideograms are all around us. For example, in the United States the silhouette of an animal with antlers on a diamond-shaped, yellow road sign is understood by all as a warning that deer could be crossing the road ahead. A cigarette symbol with a red diagonal running across it is universally understood as meaning "No smoking."

	Fish	Bird	Ax	Arrow	Bottle
MESOPOTAMIA					
EGYPT					
CHINA					

Clay seals from Mesopotamia used a series of unique symbols that represented a person's identity.

The Roman codex was a folding wax tablet which was written on with a stylus. This design led to the paper book.

was totally unrelated to mouth. In the 3rd millennium BCE, the Sumerian language had elements of polyphony—some signs had multiple phonetic values—and homophony—several signs had the same phonetic value. In fact, some syllables had more than ten symbols.

Early alphabets

The Proto-Sinaitic scripts of Middle Kingdom Egypt (1800–1500 BCE), which were inscribed on stones in the Sinai Peninsula, are the earliest examples of writing with an alphabet. Developed from ancient Egyptian hieroglyphs, they were a precursor of the Phoenician alphabet, in use in the eastern Mediterranean at the end of the 2nd millennium BCE. This had 22 consonant letters but no vowels. Merchants spread the Phoenician language around the Mediterranean region, and it was adopted and modified to create numerous other alphabets. These included Aramaic, from which Hebrew and Arabic developed, and Greek, from which Latin evolved.

Writing media

Clay tablets and stone inscriptions were not the only media for writing. Papyrus was used in ancient Egypt from at least 3000 BCE, and Chinese scribes inscribed characters onto turtle shells and animal bones. From the 1st millennium BCE, the Maya of Mesoamerica painted scripts on ceramics and walls, and carved them in wood and stone, while the Romans used wax tablets, which became commonplace at least among those with money, with bronze styluses replacing reeds. The Chinese invented paper, and the earliest known book printed on paper—dated 868 CE—was a Chinese edition of the Buddhist Diamond Sutra.

PAPER

Ancient Egyptians made the forerunner of paper, papyrus, by pounding layers of plant pith together to form uneven sheets. In the 2nd century BCE, a Chinese inventor (traditionally named as Cai Lun) devised a method for making lighter, smoother material, more akin to modern paper. A suspension of pulped tree bark and hemp waste was drained through coarse cloth and the residue that was collected was left to dry to produce paper sheets.

19 Machines

MACHINES ARE DEVICES THAT MAKE TASKS EASIER TO PERFORM. The simplest of them were used long before the dawn of civilization to move, lift, and cut objects and materials.

Machines work by changing the magnitude or direction of a force. So, a force applied to one part (the effort) moves another part to overcome a resisting force (the load). Inclined planes, wedges, levers, screws, pulleys, wheels, and axles are all examples of simple machines. Even complex modern machineries are merely collections of many of these simple ones working together.

Wedge and slope

The oldest machine was the wedge. The thin edge of a wedge pushes against an object with greater force than it takes to set the wedge in motion, so the wedge cuts through other objects. A stone ax struck against a tree, the tip of a hunter's arrow striking a deer, and a bronze knife cutting animal hide are all simple wedge machines. A wedge reorientated becomes an inclined plane, or slope. An inclined plane is a machine because it makes it easier to raise an object. The horizontal distance an object has to be moved is greater than if it was lifted vertically, but the force necessary to move it is less. The simplest example is a gentle slope. Early humans found it easier to drag building materials or animal carcasses up slopes rather than attempting to lift them vertically. And ancient Egyptians built low-gradient ramps when they were dragging heavy blocks of stone to construct pyramids.

The Archimedes screw lifts liquids or powders using a slope wrapped around an axle.

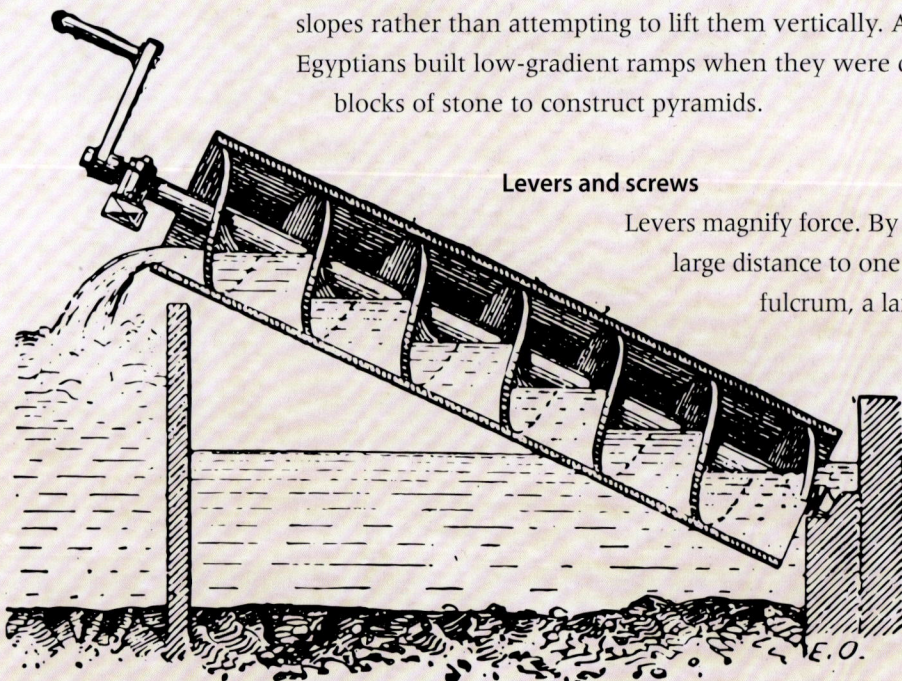

THE DIOLKOS WAGONWAY

The world's first railroad was built from coast to coast across the Isthmus of Corinth, Greece, in the 6th century BCE. The paved limestone Diolkos track is 6-8.5 km (3.7-5.3 miles) long and has parallel grooves 1.6 m (63 inches) apart along which wagons—or even boats—could be pulled. It was built to avoid having to transport cargo by boat all the way around the Peleponnese peninsula and was in use until the 1st century CE. The wagonway follows the contours of hills to keep gradients gentle, so it is a series of inclined planes.

Levers and screws

Levers magnify force. By exerting a small force over a large distance to one end of a lever pivoting on a fixed fulcrum, a large force is produced over a small distance at the other end. The earliest known scissors, simple levers, were made out of bronze by the Mesopotamians in the 2nd millennium BCE. Later, trebuchets used more complex levers to propel rocks or burning projectiles at enemies from the 4th century BCE.

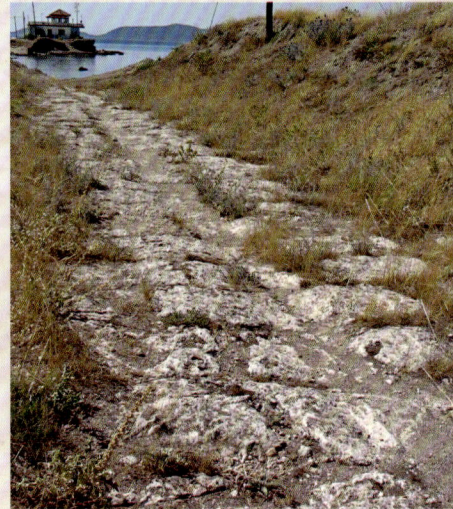

The screw, another simple machine, is first known from 5th-century BCE Greece. A screw's thread is actually a form of inclined plane wound around a central shaft. Turning the screw moves the thread, and it takes less effort to turn it into an object than it would to push it straight in.

Wheels and pulleys

In earliest times, potters rotated their clay on "slow wheels" made of stone to improve the shape of their pottery. In the late 6th century BCE, or possibly even earlier, millers used hand-rotated millstones to grind grain and make flour in Carthage. Wooden cart wheels that rotated on axles are known to have been used in the following millennium. Pulleys also employed wheels. The simplest pulley is a rope wrapped around a wheel to change the direction of a force—for example, to lift or lower an object vertically rather than pull it horizontally. In a compound pulley, the load is divided between two or more wheels so a small effort can lift a large load. The ancient Greeks invented a pulley crane in the 6th century BCE, a far more efficient way of lifting loads than using a long, inclined plane. The Romans built more complex machines, including the trispastos, a crane with three pulleys that enabled one man to lift a 150 kg (330-pound) weight.

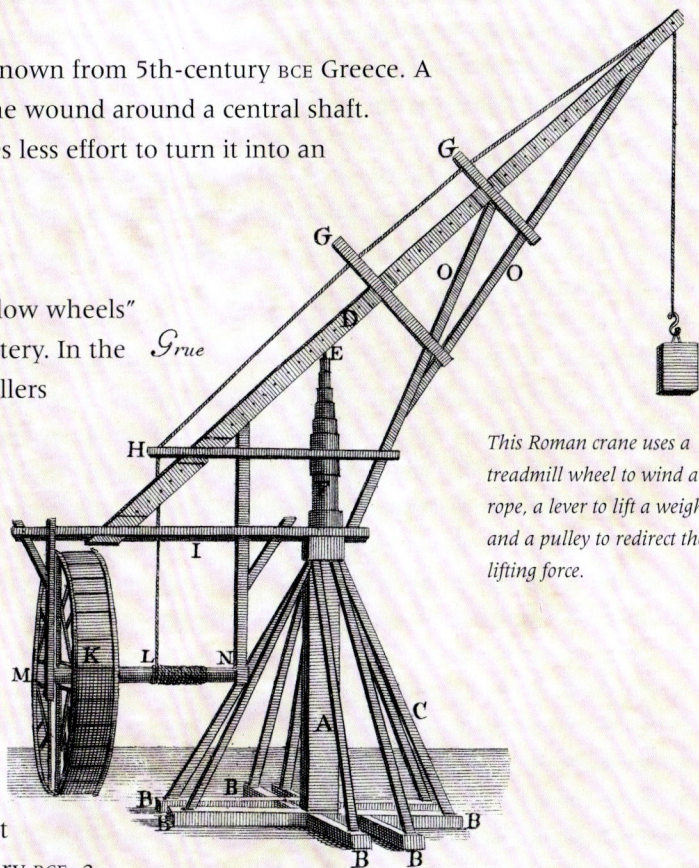

This Roman crane uses a treadmill wheel to wind a rope, a lever to lift a weight, and a pulley to redirect the lifting force.

WHEELBARROWS

A wheelbarrow combines two types of simple machine—a wheel turning on an axle, and a lever. Because the load is centered just behind the wheel (the fulcrum) and there are long handles, it is relatively easy to raise the barrow. It needs only a fraction of the energy that would be required to lift the load vertically. There may have been wheelbarrows in 1st-millennium BCE Greece, although evidence is scant. They were certainly used in 2nd-century CE Han-dynasty China where the wheel was very tall to make it easier to move loads quickly over rough ground.

20 Money

BEFORE THE FIRST COINAGE, TRANSACTIONS COULD TAKE PLACE only by barter between two people, each of whom had something the other wanted.

Money was essential for the growth of commerce. It provided an intermediary medium, enabling a seller to choose when and where he or she wanted to be a buyer. The earliest coins date from the 7th century BCE and were found in the Temple of Artemis, Ephesus. In the following century, Croesus, king of neighboring Lydia, was the first ruler to mint coins for circulation. Each coin had a fixed value, so it did not have to be weighed to establish its true worth. So began a rapid expansion in the use of money—coins spread west to Greece and east to Persia, and by the end of the 4th century BCE, they were also in circulation in Egypt and Rome.

Around the same time, in China coins were cast by pouring molten bronze into molds. They were stamped with characters, like their counterparts in civilizations to the west, but, unlike those, Chinese coins were spade- or knife-shaped. Emperor Qin Shi Huang introduced the first round coins (with a square hole in the center) in China in the late 3rd century BCE.

DOUBLE ENTRY BOOKKEEPING

The fundamental principle of accounting is that every debit must have a corresponding credit, and vice versa. This is called double-entry bookkeeping. Although the system was in use much earlier, it was not properly codified until 1494, when Italian mathematician Luca Pacioli published *Summa de arithmetica, geometrica, proportioni et proportionalita*. He was later described as the "father of accountancy" because he explained how double-entry bookkeeping should be applied consistently, explaining concepts such as balance sheets and income statements.

Paper money

The first known paper money was used in Tang dynasty China from about 800 CE. These bills were government certificates given to local merchants in lieu of transporting large quantities of coinage. The recipients could exchange them for cash later. Over time, this system evolved into a paper currency that was in general circulation. Side by side with the increase in money in circulation came an expansion of professional banking. Bankers took deposits, made loans, and changed money from one currency to another. Moneylenders accepted payment in one city and arranged credit in another, thereby removing the need to travel with large piles of coins.

21 The Lighthouse

WITHOUT ACCURATE CHARTS, EARLY SAILORS OFTEN HAD NO WAY OF KNOWING how close they were to dangerous coastlines at night. So lighthouses were built to help avoid shipwrecks.

The first aids to help mariners at night were fiery beacons on prominent hills near treacherous coasts. In the 5th century BCE, a small stone column was built near the entrance to Athens' harbor at Piraeus, probably with a fire atop to warn sailors. During the 3rd century BCE, a much more spectacular lighthouse aid was built, the Pharos of Alexandria, on the north coast of Egypt. Constructed from stone, more than 100 meters (330 feet) tall, its illumination came from a furnace at the top. The Romans took lighthouse building to a new level and erected them from the Black Sea to Dover, in southern England. Incredibly, one of them, the 55-m (180-foot) Tower of Hercules on the coast of Spain, is still operational. Early lighthouses were usually fired by wood, but sometimes coal. Later examples used window glass to keep the flames alight in gales.

The Pharos was one of the Seven Wonders of the World until destroyed by a series of earthquakes in the Middle Ages.

22 Horse Power

PEOPLE FIRST CAPTURED, TAMED, AND RODE HORSES AT LEAST 5,000 YEARS AGO, but riding equipment was rudimentary. Skilled horsemanship was only possible after a number of important innovations were made.

The saddle revolutionized horse-riding—there was less chance of falling off when riding fast! The nomadic Scythians of Central Asia used padded saddles with a pommel in the 4th century BCE, giving their warriors a big advantage over their neighbors. Stirrups, which allow a rider to lean far to the left or right, a big advantage for horseback warriors, were first portrayed on Chinese Sima Jin-dynasty artifacts dating from 322 CE. The collar harness, first used in northern China in the late 5th century, uses the horse's full strength to pull a heavy load, making it a draft animal alongside oxen.

As well as a saddle, Scythian horsemen also used bits, bridles, and reins to control their steeds.

23 The Mill

WITH AN EXPANDING HUMAN POPULATION CAME INCREASED DEMAND FOR BREAD. Consequently, flour had to be milled in ever-greater quantities, stimulating a series of revolutions in mill technology.

Round millstones found in a ship that was wrecked off Mallorca between 375 and 350 BCE are known to have originated from the Mulargia area of Sardinia, which at that time was part of the Carthaginian empire. They are the first-known examples of millstones used in a donkey- or ox-powered mill and may have been bound for Spain. Carthage was an important trading hub and an agricultural center in its own right. Hand-driven mills had been used there to grind grain and make flour since at least the late 6th century BCE. But animal power could turn larger grindstones and increase productivity dramatically. Although invented by Carthaginians, the idea of animal-driven mills caught on and rapidly spread to Sicily and Italy in the 3rd century BCE, then across the Middle East.

The first source of power used to drive mills was the human body, as seen here on a treadmill. Later, animals were used to drive the machines.

GEARS IN ANCIENT CHINA

The first gears were probably invented in China in the 4th century BCE, and there is no doubt that the Romans used them in their watermills three centuries later because the contemporary engineer Vitruvius described them. Gear technology relied on a series of connected toothed wheels and axles to transform the energy of a rotating waterwheel into energy to turn machinery. Contrary to the practice of later centuries, Vitruvius described Roman mills using step-down gearing, with the grindstones turning more slowly than the water wheel. Later, step-up gearing was usual. For centuries, gear systems were made from precision-crafted wood.

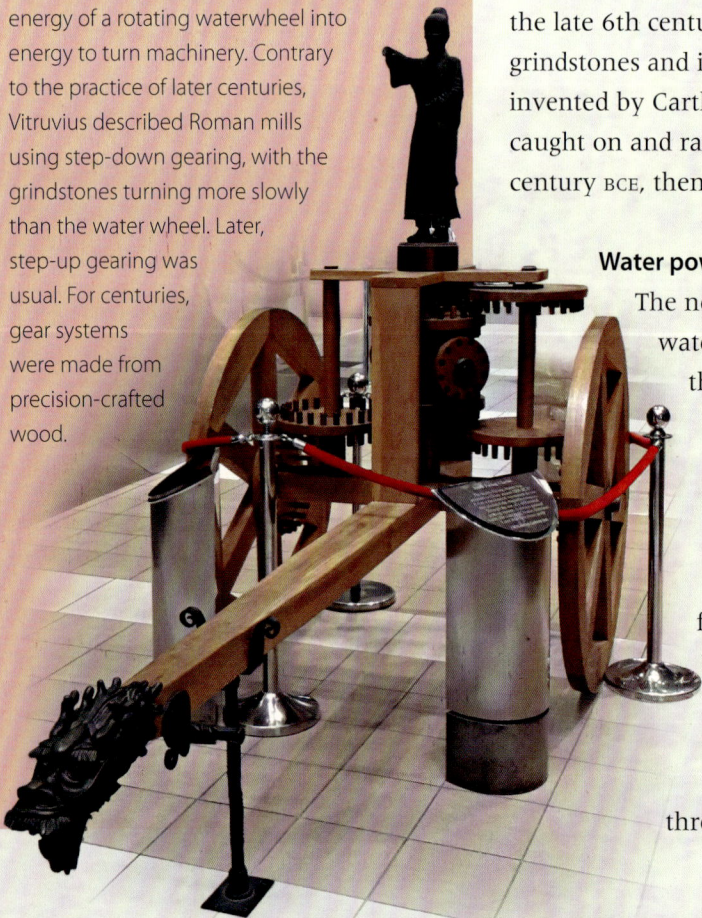

Water power

The next big technological advance was the watermill. It probably evolved from a device called the noria, a waterwheel with scoops to lift water to a higher level in Egyptian irrigation schemes. Water power removed the need for human or animal labor and increased a mill's productivity dramatically. In theory, the mill could grind flour as long as the stream flowed. In the 3rd century BCE, Greek engineer Philo of Byzantium wrote about a watermill at Perachora in the Peloponnese. The first examples had paddles that rotated in a horizontal plane and a vertical shaft connected through a hole in the stationary grindstone to

Left: A south-pointing chariot, an early form of compass used in China, used wooden gear wheels.

WINDMILLS

Wind-powered machines were first used to grind grain in Persia at some point between the 7th and 9th centuries. Called panemone windmills (left), they had vertical sails that swung around in the wind and turned a central, upright axle. The axle drove grindstones attached at the base. In China, windmills were used for milling and pumping water in the 12th century. Northwest Europe's first windmills were built in the 1180s. Called post mills, all the mill machinery was mounted on a single vertical post, around which it could be turned to bring the sails into the wind. Early examples were unstable and many blew over in strong winds. Over time, more robust designs were invented, including the open trestle post mills built in England, France, Germany, Belgium, the Netherlands, and New England.

the turning upper stone. As the flowing water of a stream pushed the paddles, so the wheel, axle, and stone rotated, grinding the grain.

Vertical waterwheel

Horizontal waterwheels were later replaced by more sophisticated—and more efficient—vertical undershot wheels. (Undershot means the flowing of the water contacts the bottom of the wheel.) The force of flowing water pushed the wheel paddles, turning the wheel, and power was transmitted to grinding millstones via an axle and interconnected toothed wheels called gears.

The invention of the overshot wheel led to another increase in productivity. Now, water was directed from a trough to paddles at the top of the wheel, causing it to turn. The Romans built a mill complex at Barbegal, in southern France, which employed 16 overshot wheels arranged in two parallel sets of eight to produce 4 metric tons (4.5 tons) of flour daily in the 2nd and 3rd centuries BCE. Tide mills were built in some coastal regions where there was a sufficient tidal range to make them viable, the first being a 6th-century, vertical-wheeled example near Waterford, Ireland. According to the Domesday Book, in 1086 there were 5,624 watermills in England alone. Apart from the invention of the windmill, mill technology changed little until the advent of steam power in the late 18th century.

This proposed mechanism from the 1880s uses the power of the horse (walking on a treadmill) to not only drive a flour mill, but also to clean the stall automatically, cut up the fodder finely, and even pump water for extinguishing fires.

Left: A model of an undershot watermill based on the first technical description of one written by Roman engineer Vitruvius in the 1st century BCE.

24 The Arch

AN ARCH IS CONSIDERABLY STRONGER THAN A LINTEL—A HORIZONTAL BAR—OF THE SAME WIDTH. Whereas lintels will crack under heavy loads, curved arches direct the weight into their supports.

The invention of the arch would revolutionize architecture because it allowed the use of larger doors, windows, and bridges.

The earliest known curved arch (c.2000 BCE) is made of mud bricks and spans a doorway at the Mesopotamian site of Tell Taya, Iraq. Another, the city gate of the Canaanite city of Ashkelon, Israel, dates from around 1850 BCE. Neither of these had the strength of later arches, however, because they didn't use wedge-shaped stones called voussoirs. Voussoirs help spread the weight above an arch more effectively, since the downward pressure pushes them together, rather than apart. There is an early example, dating from the 4th century BCE, in the Greek city of Rhodes. Clearly, the ancient Greeks knew about arch technology, but they did not use it very much. Neither did the ancient Egyptians.

Roman innovation

It was the Romans who realized the true potential of this architectural device. They used voussoirs supported by strong columns. The columns buttress the downward force, or thrust, and conduct it into the foundations. As an arch was being constructed, Roman builders supported it with wooden scaffolding until the final piece—the keystone at the very top—was put in place. The Romans incorporated thousands of semicircular arches into their building program throughout the empire to support bridges and aqueducts, and in gateways, public buildings, and villas. To celebrate military victories, they constructed triumphal arches. There were once more than 50 of those in Rome alone. In that city's magnificent Colosseum arena there are 80 arches on each floor. Pointed, rather than semicircular, arches had been used in India in temples and palaces dating back to the 3rd century BCE, but they reached their zenith in the Gothic style of church construction in the Middle Ages in Europe.

The arch's enormous strength means it is still used in modern structures. For example, the Hoover Dam, built in the 1930s, is an arch on its side, able to withstand the immense force of the Colorado River.

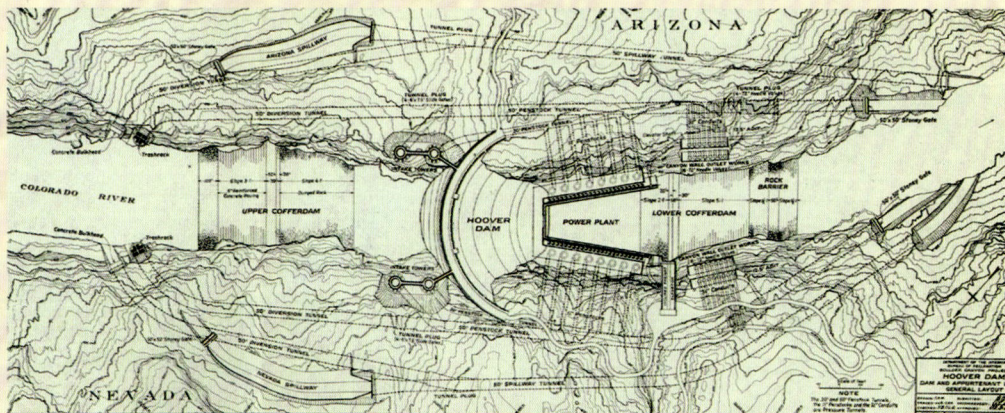

25 Heron's Inventions

HERON OF ALEXANDRIA WAS A 1ST-CENTURY CE INVENTOR WHO IMAGINED AN AUTOMATED WORLD, which we are still striving toward 2,000 years later. His most famous creation was the aeolipile, the world's first steam engine.

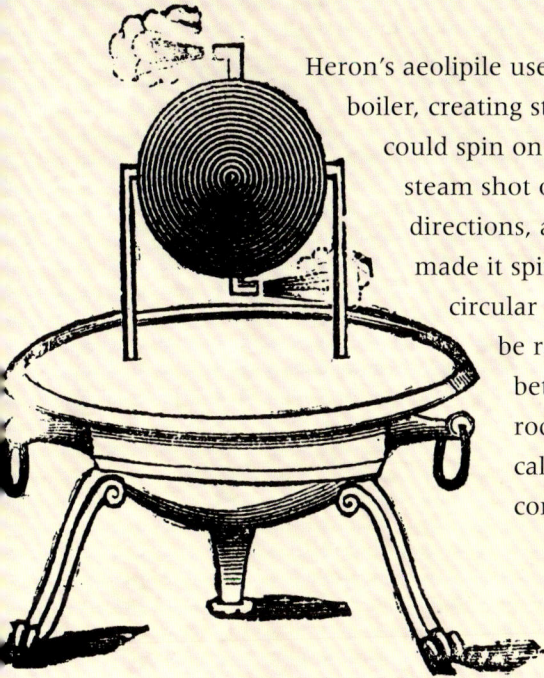

Heron's aeolipile used steam power to create motion. Fire heated water in a boiler, creating steam that was piped into a copper sphere. The sphere, which could spin on its axis, had two open-ended nozzles, through which the steam shot out at great speed. Since the nozzles were bent in opposite directions, and were on opposite sides of the sphere, the steam exhaust made it spin. Another Heron device is the dioptra. This consisted of a circular plate, calibrated with angles, on a stand. The plate could be rotated and tipped to allow a surveyor to measure the angle between different objects, near or far, with the aid of a sighting rod. People could triangulate to produce accurate plans or calculate the position of astronomical bodies, so the dioptra is considered an early theodolite.

No one knows what practical uses Heron imagined for his steam engine. Even if he thought of it as no more than a fascinating experiment, it certainly influenced later inventors.

Robots and vending machines

Heron also described a syringe, an automatic door-opener, and a water fountain. Unlike his dioptra, some of his inventions may not have advanced beyond sketches and descriptions. But he did build a programmable robot, which was a cart powered by a falling weight, its program consisting of string wrapped around the drive axle. Using pegs projecting from the axle, Heron could vary how the rope was wound around it, allowing the robot to change direction and move along a pre-programmed course. The peg-and-rope mechanism is reminiscent of the punch cards used in 20th-century computers (see more page 62).

Another creation of Heron's was the world's first vending machine, which dispensed holy water. If a coin was popped into a slot in the machine it fell onto a tray connected to a lever. The tray tilted, tipping the coin off, and the lever opened a valve, which dispensed the water. Heron also designed a hand-operated water pump, which the Romans later used with a fire hose to extinguish blazes. Had this pump been combined with his steam engine, it could have resulted in a powerful steam-driven water pump, but no one put the two together, at least not until the early 18th century.

Heron designed a musical organ that was powered by the wind. The windmill powered a piston which forced air through the pipes, making music.

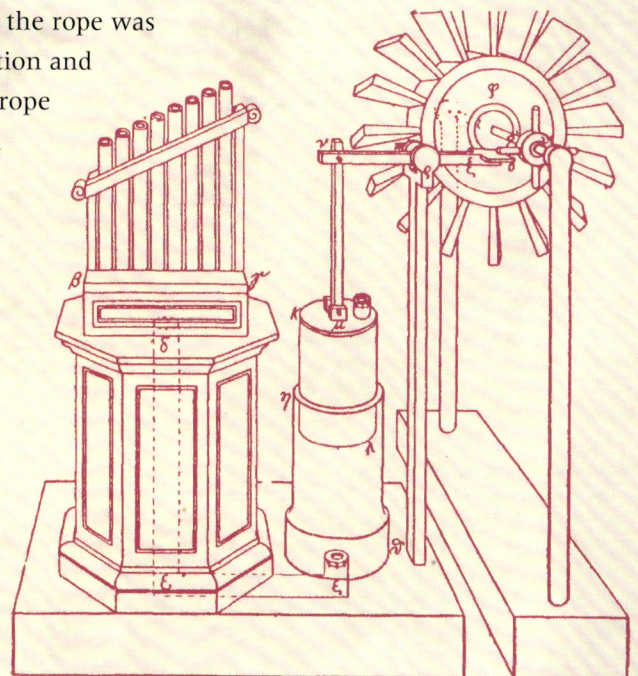

26 The Seismometer

EARTHQUAKES ARE ONE OF THE MOST DESTRUCTIVE FORCES IN NATURE, even capable of collapsing a civilization. In 132 CE, a Chinese poet invented an early warning system.

Zhang Heng's "earthquake weather vane" could pick up tremors from all directions. The ball dropped into the frog closest to the earthquake's epicenter.

The inventor was Zhang Heng, whose extraordinary contributions also included the Chinese solar calendar and a primitive compass system. Zhang Heng's earthquake detector is said to be the first seismograph (from the Greek for "earthquake writer"). It was a bronze urn with a pendulum inside, which swung when small, even imperceptible, tremors spread through the ground. This movement dislodged one of eight balls held in dragon heads equally spaced around the device, alerting the royal court to an earthquake. This allowed officials to prepare for bigger quakes to come and also gave early warning that a more distant province had been struck and needed help.

27 Explosives

SALTPETER HAD BEEN USED IN CHINESE MEDICAL CONCOCTIONS FOR CENTURIES AS IT WAS THOUGHT TO PROLONG LIFE. In combination with charcoal and sulfur, it proved to be quite an effective way of ending it!

Sometime in the 9th century an inquiring Chinese alchemist experimented with a mix of potassium nitrate (saltpeter), charcoal, and sulfur in proportions of approximately 75:15:10 by weight. The results when this mixture was exposed to a flame were doubtless startling. A contemporary account describes "hands and faces burned and the whole house where they were working burned down." The armies of the Southern Wu began using gunpowder devices against the Mongols as early as 904 CE. Devices such as "flying fire,"

CHINESE FIREWORKS

Huo yao, the "fire chemical," was used for celebrations before it was used in warfare. Bamboo tubes filled with gunpowder were thrown onto fires during festivals so that the noise might scare away evil spirits. Franciscan monk Roger Bacon, one of the first to experiment with gunpowder in Europe, wrote in 1242: "… if you light it you will get thunder and lightning if you know the trick." The displays we enjoy today are largely thanks to the Italians, who incorporated trace amounts of metals to their gunpowder mix in the 1830s to produce colorful explosions.

Greek Fire was an incendiary weapon used in the Middle Ages. While its purpose was obvious—it set fire to the enemy—its constituents remain a mystery. The best guess is it is a mixture of naphtha (a paraffin wax) and quicklime. On contact with water, the lime generates heat and sets fire to the naphtha.

an arrow with a burning tube of gunpowder attached to the shaft, rudimentary hand grenades, fire lances, a type of primitive flamethrower, and landmines were all deployed. The first bombs were used in the siege of Qizhou in 1221. Iron cannisters thrown by catapults showered deadly fragments of jagged metal when they exploded.

Worldwide weapon

Despite Chinese efforts to keep it to themselves, knowledge of gunpowder eventually spread to the rest of the world. The Mongols brought gunpowder with them as they conquered Eurasia and, by 1350, English and French armies were facing off with cannons. A major improvement came when 15th-century Europeans added water to the mix to form a paste that could be dried and ground into "corned" powder, which was more durable and reliable, ensuring that the ingredients would ignite at the same time. Medieval gunners favored large grains for cannon, medium grains for shoulder arms, and fine grains for pistols.

DETONATION

Relatively low-energy explosives like gunpowder rely on a rapid combustion of their constituents to generate energy—in this case charcoal and sulfur are the fuel and saltpeter is the oxidizing agent. Higher-energy explosives, like dynamite or TNT, involve a breakdown of the molecules that make them up. They don't need oxygen or any other reactant, but high temperatures aren't enough to set them off. What they need is a trigger. Dynamite is detonated by igniting a small cap of low-explosive gunpowder. This sends a shockwave through the dynamite that triggers the rapid breakdown of its chemical bonds, and the release of large amounts of destructive energy.

After 1867 the Swedish chemist Alfred Nobel became one of the world's richest men with explosives factories across the world. His empire was built on the invention of dynamite that year, which combined highly unstable nitroglycerine with a soft, powdered earth to make a high explosive that could be handled safely.

28 Printing

WRITING WAS A HUGE STEP FORWARD FOR HUMAN CIVILIZATION, but it took time to spread the written word to a wide audience. The first books were copied out by hand and available only to the wealthy and powerful few.

LEONARDVS VINTIVS
FLORENT. PICTOR.

Laudii Apelleæ metuit fua fidera fulgor.
Soffite me vinci, me moriente mox
VAL. THILO L.
ANDREAS

3062·4

WOODBLOCK PRINTING

One of the earliest surviving books is a Buddhist text known as "The Diamond Sutra." It was created in 868 in China using block printing, in which text handwritten on paper is glued to a wooden block which is then carved out using a special knife, leaving the text standing proud. The block is then inked and the print made.

In his enticingly named *Dream Pool Essays* of 1086, the Chinese scientist Shen Kuo described the invention made by a peasant alchemist called Bi Sheng. In 1041, Sheng made a series of clay tablets, inscribing a Chinese character on the surface of each before firing them. In doing so he invented type. Sheng's tablets could be set up fairly quickly, but the biggest impediment to their success was the Chinese language itself, which employed over 5,000 distinct characters. Also, the clay tablets weren't very durable.

Wang Zhen, a 14th-century Chinese government official, created a set of movable characters using wood. He also devised a revolving table that allowed the typesetter to access the wooden characters more easily, but the wooden blocks deteriorated with use and producing new ones was time consuming. More durable metal type was developed in Korea in the late 14th century. A Korean monk named Baegun is credited with printing a Buddhist text using movable metal type in 1377.

THE GUTENBERG PRESS

In 1450, German Johannes Gutenberg developed a technique of arranging letters cast from metal molds to produce blocks of text for a whole page. His printing press was adapted from grape presses used by winemakers. Hundreds of copies of one page could be printed, and the blocks could easily be broken up and the letters reused.

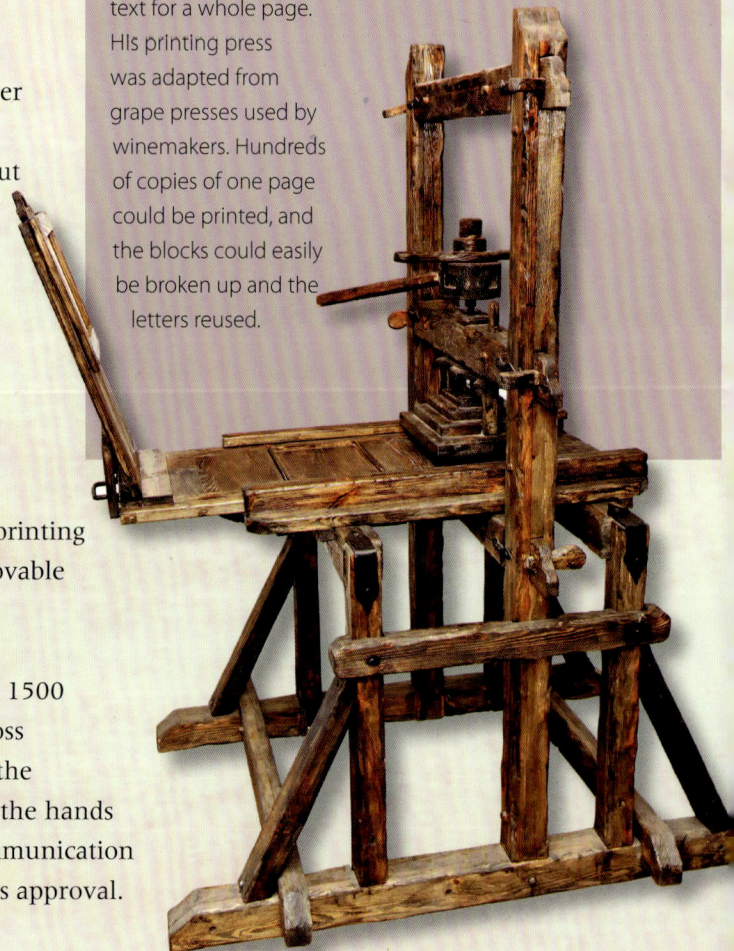

The movable metal type press invented by Gutenberg (see box) was a runaway success. By 1500 there were over 1,000 printers in operation across Europe. In 1501, Pope Alexander V recognized the threat to his power if wider knowledge fell into the hands of the general population. He threatened excommunication for anyone who printed manuscripts without his approval.

29 Navigation

For most of human history people navigated through the landscape using familiar landmarks and the position of heavenly bodies. In the 12th century, compass technology began a revolution in getting from A to B.

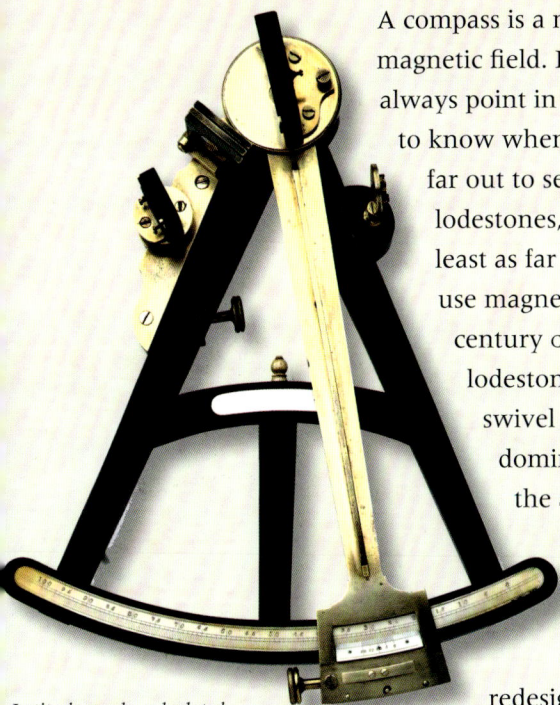

A compass is a magnetic device that aligns with Earth's own magnetic field. Left free to move, the compass will therefore always point in the same direction, which is a useful thing to know when you are lost in a strange, unknown land (or far out to sea). Naturally magnetic iron ores, known as lodestones, were being investigated by ancient cultures at least as far back as 600 BCE. The Chinese were the first to use magnetic phenomena as a practical tool. By the 1st century CE, they were employing "southern governors," lodestones carved into spoon shapes that were free to swivel on a flat board. While today, the north is the dominant direction from which others are measured, the ancient Chinese saw things differently. These early compasses were not used for navigation but for divining the most auspicious orientations for temples and the like. It took another thousand years before the compass was redesigned from southern governor to "iron fish." In this innovation a wafer-thin metal pointer floating on a dish of water swung around to point to the south. This wet compass was suitable for use aboard ships, and the technology allowed navigators to cross the seas with greater confidence—and in turn spread compass tech farther and wider. The first record of the magnetic compass in Europe is in Amalfi, Italy, at the beginning of the 14th century. It is possible that it made the decisive contribution to the rise of the Italian city-states based on maritime trading routes.

Latitude can be calculated by the angle of the Sun—it is higher in the sky when in tropical regions than when nearer to the poles. The measurement was made by aligning a sight to the Sun and reading the angle from a graded arc. The peak of innovation was the sextant, designed by John Hadley around 1731. It used mirrors so the navigator could see the Sun and the horizon at the same time.

MARINE CHRONOMETERS

The Earth rotates west to east, turning once every 24 hours. That creates an apparent motion of the Sun through the sky, reaching the highest point each day at noon. Places with the same longitude share the same noon. Places to the east have earlier noons, and midday comes later to the west. Comparing the local noon to the time at a fixed location tells you your longitude (every hour of difference is a 15° shift). However, theory ran far ahead of technology, because it took until the late 18th century before clocks, known as marine chronometers, became accurate enough for the task.

A nautical compass points north, and all other directions are measured as angles, or bearings, from that mark.

Global positioning

Direction is not the only thing a navigator needs to know. Voyages out of sight from land relied on dead reckoning: estimating the current position of the ship based on how long and how fast it had been traveling. This was a hit and miss affair to say the least—often hit, when vessels strayed way off course into dangerous waters. The solution was to measure the latitude (position north or south of the Equator, which involved measuring the angle of the Sun), and the longitude (position east or west). This was a trickier task but was solved using clocks (see box, above).

30 Guns

IN THE SIMPLEST TERMS, A GUN IS A TUBE FROM WHICH A PROJECTILE IS FIRED BY MEANS OF A PROPELLANT. Over time, the gun grew more reliable, deadlier, and played a major role in the course of human history and warfare.

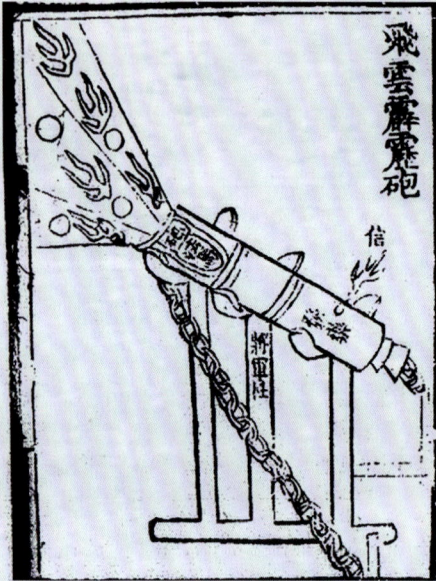

Chinese fire lances made from bamboo, some of the first weapons to make use of gunpowder, were the forerunners of the gun. When the gunpowder was ignited at one end, arrows, shards of pottery, or metal fragments were propelled toward the enemy. Around the 1100s, metal tubes began to replace the bamboo ones, and shrapnel was replaced by shaped projectiles. A carving of 1128 may be the first depiction of these "eruptors," which were able to throw an iron ball about 40 meters (131 feet). An actual example of one of these early guns, dating from 1288, suggests that they were braced against the ground by means of a wooden pole, which dealt with the problem of recoil and also helped distance the soldier from the flames shooting from the barrel.

A musketeer from the 1600s.

A Ming-dynasty "eruptor."

Handheld weapons

Gunpowder weaponry soon spread beyond China. By the 14th-century bigger and better cannons in Europe spelled the death knell for fighting forces organized around knights in armor, but they were unwieldy and soldiers wanted guns they could carry. The simple "hand gonnes" were basically miniature cannons that could be fired over the shoulder. Often it would take two men to wield the hand gonne: one to aim and one to fire.

Lock and stock

The mid-fifteenth century saw the advent of the arquebus, the first gun designed with a stock that allowed it to be fired from the shoulder, and later

RIFLES

Sometime in the 15th century, gunmakers discovered that a gun with a rifled bore (shallow grooves cut inside the barrel) made bullets spin, and that spinning bullets had a greater range and were more accurate. Originally a single-shot, muzzle-loading gun, the rifle has since developed into a potent weapon capable of firing multiple rounds over a range of 1,800 meters (5900 feet).

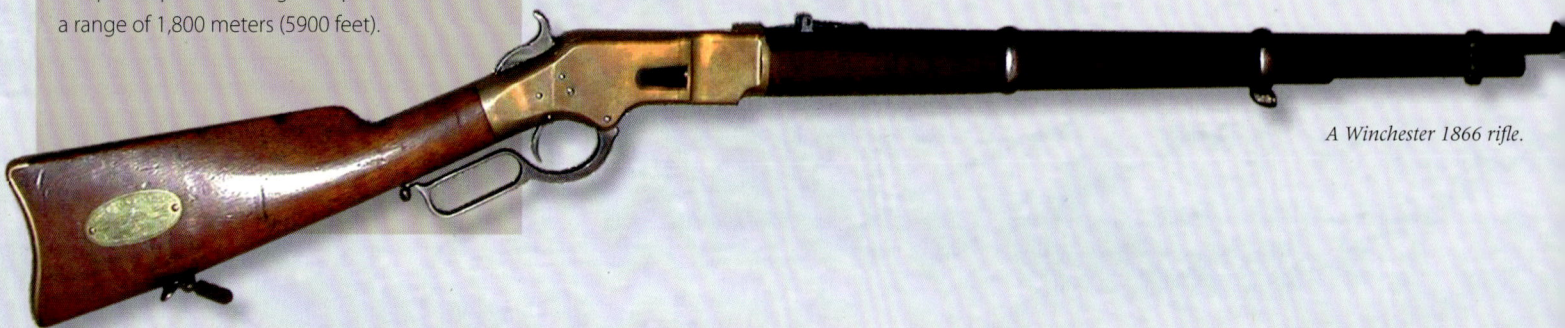

A Winchester 1866 rifle.

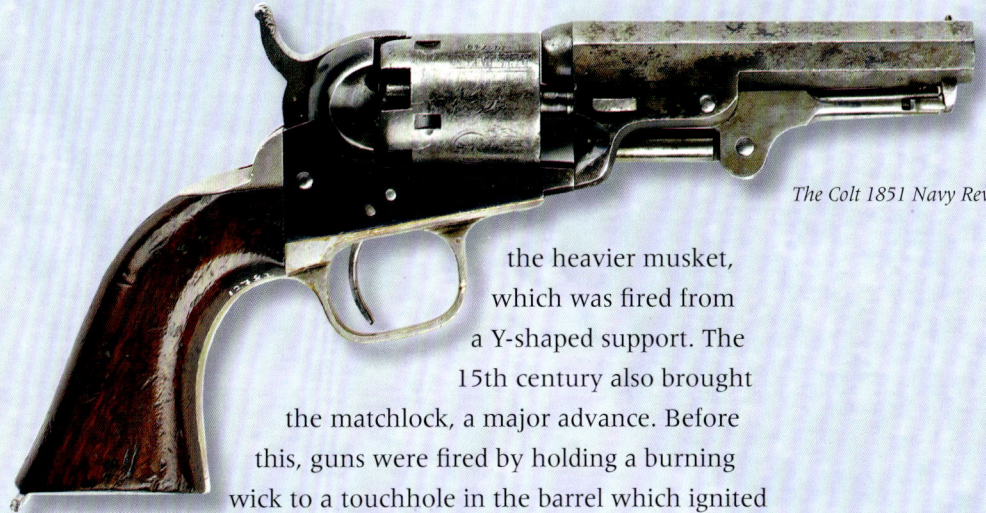
The Colt 1851 Navy Revolver.

THE MAXIM GUN

In 1884, American-born British inventor Hiram Maxim invented the first fully automatic machine gun. It harnessed the recoil energy from each bullet fired to eject the used cartridge and pull in the next one. It had a rate of fire of 600 rounds per minute and was cooled by a water jacket surrounding the barrel. It became a mainstay of the British and later the Austrian, German, Italian, Swiss, and Russian armies, and featured heavily in the carnage of World War I. The Maxim gun also gave European nations military superiority in the colonization of Africa.

the heavier musket, which was fired from a Y-shaped support. The 15th century also brought the matchlock, a major advance. Before this, guns were fired by holding a burning wick to a touchhole in the barrel which ignited the powder. The matchlock held the powder in a flash pan, ignited by a wick, or match, in a movable clamp. Both the gunner's hands could remain on the gun, vastly improving aim. The later wheel lock generated a spark mechanically, and was even more reliable as there was no wick to be kept lit.

The flintlock, most likely invented by Marin le Bourgeoys in France in the 17th century, was the first gun to have a trigger mechanism. When it was pulled, this mechanism caused a hammer, called a frizzen, to strike a flint and shower sparks into gunpowder in a priming pan. The flintlock could even be used on horseback!

31 The Steam Turbine

THE IDEA THAT STEAM COULD BE MADE TO DO USEFUL WORK IS OFTEN ASSOCIATED WITH THE STEAM ENGINES of the Industrial Revolution. In fact, the idea of the steam engine had been invented far earlier.

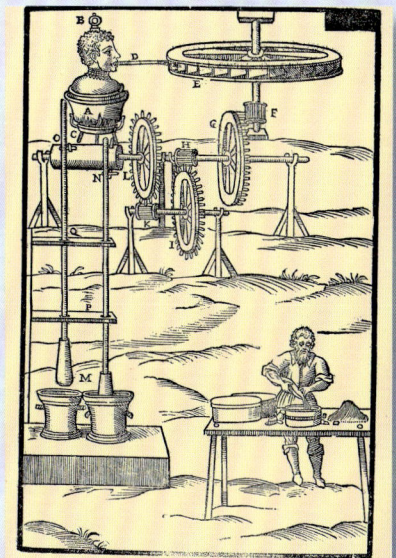

In 1551, Taqi ad-Din, a polymath from Damascus who worked throughout the Ottoman Empire, recorded a turbine being used to turn a cooking spit. The smoke and steam from the fire pushed against fan-like vanes attached to the axle of the spit, thus creating the first practical (although limited) engine. Later steam engines concentrated more on using vapor to drive pistons to turn wheels. It was only in 1884 that the steam turbine was reinvigorated by Charles Parsons. Hot steam flows past the fan-like blades of the turbine, converting a linear flow into a rotational motion. Even today, spinning steam turbines generate 75 percent of the world's electricity.

A 1629 design for a steam-powered device for crushing rocks.

32 Maps

PEOPLE HAVE LONG RECOGNIZED THE VALUE OF A MAP. It's good to know where you are in the world, and in which direction to go if you want to get to somewhere else.

In 2009, a 14,000-year-old stone tablet was found in a cave in Spain. Scratched on it is what may be a map of the area around the cave, showing features such as a river, a mountain, and good places to hunt deer. The Imago Mundi, the oldest known map of the world, or at least the world as it was known to the Babylonians, dates back to around the 6th century BCE. A clay tablet shows Babylon surrounded by other cities, mountains, and water.

In the 2nd century CE, the astronomer and astrologer Claudius Ptolemy realized that, in order to draw up an accurate horoscope, he needed to determine a person's precise place of birth. He devised a system of lines of latitude and longitude and plotted some 10,000 locations from Britain to Europe, Asia, and North Africa on his map of the world. Ptolemy knew the Earth was round and proposed ways of representing this on a flat surface. He published his findings in *Guide to Geography*, an eight-book work, that provided a description of methods for mapping the known world.

Accurate mapping was largely lost in Europe during the medieval period. The famous world maps or "Mappa Mundi" of the time were more religious representations of the Earth than scientific ones. This changed in the 15th century when explorers and traders began to spread out across the world and needed to chart their courses. The invention of the printing press in the middle of the century also meant that maps could be made widely available.

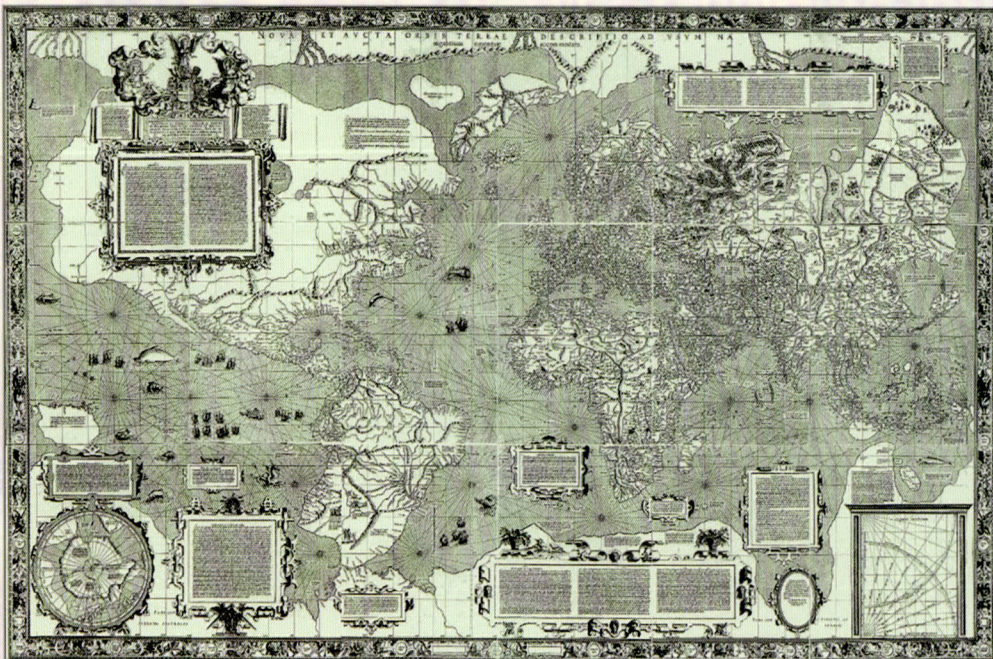

A major problem with making a world map is that the world is round and a map is flat. The lines of latitude and longitude, marking out a globe, become distorted on a flat map. In 1569, Flemish geographer Gerardus Mercator created a map that sought to account for this by keeping the lines straight and gradually widening landmasses closer to the poles. The result was this Mercator projection, which was useful for navigators, but it also distorts the way we see the world, with countries in the north, like Canada and Russia, looking very big, while equatorial Africa looks much smaller than it really is.

33 Telescopes

THE TELESCOPE TRANSFORMED THE WAY WE SEE THINGS, but the invention that brought the furthest reaches of the Universe closer can't shed light on its own beginnings. The inventor of the telescope remains a mystery.

The first person to apply for a patent for "perspective glasses" was Dutch lensmaker Hans Lippershey in 1608. It is claimed that he stole the idea from Zacharias Janssen, a rival lensmaker. Jacob Metius also claimed rights to the invention, but the Dutch government turned both down, declaring the telescope so easy to reproduce it wasn't worth patenting.

Re-engineered

Lippershey's instrument was a modest one with a magnification of about three times. Galileo Galilei soon heard word of the Dutch invention and in 1609 set about designing his own. His instrument was a great improvement, magnifying objects 20 times. He pointed his telescope at the sky, revealing for the first time the mountains and craters of the Moon, four of Jupiter's moons, and the phases of Venus.

Refracting telescopes like Galileo's used a concave eyepiece lens and a convex objective lens to bend, or refract, light. In 1611, Johannes Kepler suggested making a telescope using only convex lenses. The advantage was a larger field of view and higher magnification, though the image would be upside down. Kepler's design was first used by Christoph Scheiner in the 1620s for a study of sunspots. An upside-down image was no disadvantage to an astronomer, and the astronomical telescope was generally accepted by the middle of the 17th century. The real problem for a telescope maker was in finding glass that was of sufficient quality. Images were often blurry or surrounded by colored haloes, called chromatic aberration.

In 1668, Isaac Newton declared his belief that refracting telescopes would always be flawed and presented his idea for a new kind of telescope. His reflecting telescope gathered light by means of a curved mirror, which was then directed to the eyepiece by an angled secondary mirror. The largest optical telescopes in operation today are all reflecting telescopes.

Legend has it that Hans Lippershey was inspired by watching his children playing with lenses.

SPECTACLES

Lenses for correcting vision were around a long time before telescopes and microscopes. Salvino D'Armate often gets the credit for inventing spectacles in 1284, but Marco Polo reported seeing Chinese people using them in 1270, and they in turn credited 11th-century Arabs with the idea. Even earlier, in the 1st century CE, the Roman philosopher Seneca is said to have used a glass sphere full of water to magnify his reading material.

Left: Cardinal Don Fernando Niño de Guevara is pictured in the late 16th century sporting his spectacles.

34 Microscopes

As well as being used to magnify distant objects with telescopes, the lensmaker's art was also put to work in investigating the very small. The first microscopists began to unveil the world of the invisibly small.

The same inventors who claimed the telescope are also implicated in the invention of the microscope—the likes of Hans Lippershey and Zacharias Janssen. The early microscopes were compound devices, using at least two lenses. The objective lens is positioned close to the object and produces an image that is magnified further by the second lens, called the eyepiece. To begin with, they were capable of magnifying nine times the true size. Galileo made improvements to the microscope by adding a focusing mechanism in 1609, calling it an *occhiolino*, or "little eye."

Microbiology

Even using the modest magnifying power available, biologists discovered details of the structure of insects and other living things that had previously escaped notice. In 1665, English scientist Robert Hooke published *Micrographia*, which had astonishingly detailed illustrations of natural objects like insects. Hooke sometimes had problems getting his subjects to stay still and described how he sedated a troublesome ant with brandy which "knock'd him down dead drunk, so that he became moveless." (After an hour the ant "suddenly reviv'd and ran away.")

Dutch trader Antonie van Leeuwenhoek first became interested in microscopy as a means of checking the quality of the textiles he sold. Dissatisfied with the instruments available to him, van Leeuwenhoek taught himself the art of lensmaking and began designing his own microscopes. He is known to have made more than 500, only 10 of which still survive. Basically, these were very powerful magnifying glasses, with a single lens capable of giving a magnification of more than 200 times. In common with his fellow microscope enthusiasts, van Leeuwenhoek was soon turning his lenses toward the natural world. He studied yeast, plaque from his teeth, blood and sperm cells, and the organisms he found in drops of water.

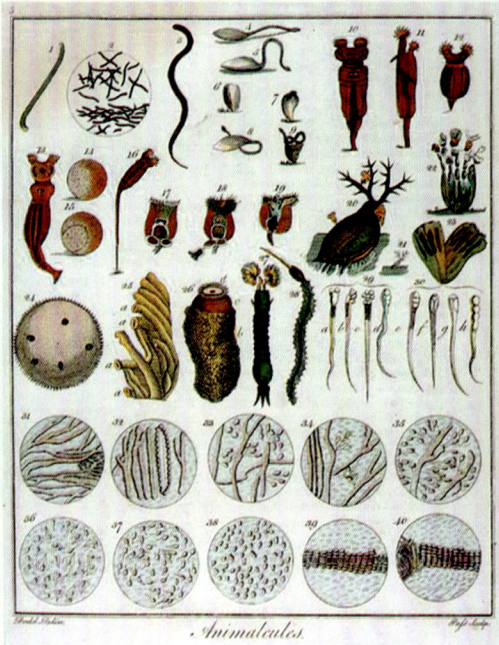

Animalcules.

In 1683, Antonie van Leeuwenhoek used his microscope design to make the first observations ever recorded of microorganisms. He called them "animalcules."

ELECTRON MICROSCOPES

The electron microscope, invented in 1931 by Ernst Ruska, uses magnets to focus electron beams with a wavelength 100,000 times shorter than that of visible light. The microscope can achieve a resolution of objects millions of times smaller than can be seen with a light microscope. The first commercially available electron microscopes appeared in 1939.

35 Calculators

IT'S IMPOSSIBLE TO IMAGINE A CIVILIZATION WITHOUT NUMBERS. The more complex the civilization, the more numbers have to be crunched to keep things organized and the greater the necessity of finding a way of doing it.

Although Blaise Pascal built more than 50 of his adding machines, his invention was a commercial failure because the devices could not be built with sufficient precision to be reliable. Another drawback was that the Pascaline could add, but not subtract.

The abacus, which most likely had its origins in Babylon 5,000 years ago, was the earliest recorded calculating device. More of a counting machine than a true calculator, it can still be found in use in East Asia today.

In 1614, eccentric Scotsman John Napier invented logarithms, a table of numbers that allowed tricky multiplication to be performed as addition. Napier also invented Napier's Bones, where the logarithm values were carved on ivory sticks which were used for calculating. The German academic William Schickard invented the "calculating clock," composed of rotating rods, representing columns of numbers, and a set of gears to move the rods and display the results. The machine, built in 1623, could perform addition and subtraction using six-digit numbers. Adding Napier's Bones to the machine meant it could be used for multiplication and division as well. Unfortunately, knowledge of Schickard's invention was lost after his death from the plague in 1635.

Mechanical math

In 1642, French mathematician Blaise Pascal invented a mechanical adding machine to assist his father's work as a tax collector. The "Pascaline" was a digital calculator in which a series of toothed gears were connected so that numbers could be added or subtracted by moving the wheels. German mathematician Gottfried von Leibniz improved Pascal's designs to produce a machine that could multiply. Von Leibniz's machine, called a step reckoner, consisted of a setup mechanism to enter the digits of the quantity to be multiplied, a handle to crank for each digit of the multiplier, a system of gears, and a result register.

POCKET CALCULATORS

The first handheld electronic calculator was built by Texas Instruments in 1967. Capable of performing arithmetic functions on 6-digit numbers, it was about the size of a large paperback, so was not really pocket-sized. The first true pocket calculator was produced by Busicom in Japan in 1970. It was the first to use an LED display and have an integrated circuit designed specifically for a calculator.

36 The Barometer

THE INVENTION OF THE BAROMETER, AN INSTRUMENT FOR MEASURING AIR PRESSURE, gave scientists a device to predict weather and measure altitude. It is generally credited to the Italian physicist Evangelista Torricelli.

In an experiment carried out around 1644, Torricelli filled a glass tube with mercury, sealed it at one end, and inverted it in a small basin which was also filled with mercury. He observed that, rather than run out into the basin, the column of mercury in the tube maintained a height (about 76 cm; 30 inches) leaving an empty gap at the top. The level of the mercury column rose and fell, often in line with changes in weather conditions. The scientific community was uncertain about what kept the mercury in place. Was it the gap pulling it up, or was it the air pushing from outside? The Frenchman Blaise Pascal showed it was the latter, the weight of the air, better known as air pressure. He had a barometer carried to a mountaintop, where the mercury dropped as the air pressure fell. That discovery helped unpick the mysteries of gases.

Pascal's brother-in-law did all the hard work to discover air pressure in 1648, while the scientist himself stayed in Paris.

37 Von Guericke's Inventions

OTTO VON GUERICKE, BÜRGERMEISTER OF MAGDEBURG AND MAGISTRATE OF BRANDENBURG, invented a vacuum pump and an electricity generator. Both these devices were used by a later generation of scientists and engineers to transform the world.

Von Guericke's electrostatic generator used the friction of the hand to create a build-up of electric charge on the ball of sulfur.

After seeing Magdeburg devastated by fire in 1631, von Guericke looked for ways to improve water pumps. In 1650, he discovered that he could use a pump to remove air from a sealed metal or glass jar, and so he invented the world's first vacuum pump. He demonstrated how his pump could be used to determine the weight of air in a vessel by comparing it when full and when sucked empty. He also discovered that light can travel through a vacuum, but sound does not.

The Magdeburg spheres

In 1654, von Guericke carried out a more famous experiment. He had two hollow bronze hemispheres made which could be joined together to form a sphere about 36 cm (14 in) across. He then used his vacuum pump to suck the air out of it through a valve in one of the hemispheres. In a convincing demonstration of the power of air pressure, a team of eight horses attached to each hemisphere failed to pull them apart. As soon as the valve was opened and air allowed back into the sphere, the two halves simply fell apart. Von Guericke reasoned that the horses could not pull the hemispheres apart because the "weight of the sky" was pushing them together. He believed that, as the sky had a finite weight, a finite amount of air surrounded the Earth, and that, at a certain height, this air would fade away.

HAUKSBEE GENERATOR

In 1706, Francis Hauksbee brought together von Guericke's inventions, using a vacuum pump to create an empty glass bulb which he used in place of the sulfur ball to collect an electric charge. Later, Stephen Gray famously suspended an eight-year-old boy from a ceiling using silk threads which acted as insulators, and charged him using a Hauksbee generator. Gray then proceeded to reveal to an enthralled audience some of the effects of electricity, for example, surprising a volunteer with an electric shock when a spark jumped from the boy's hand to his.

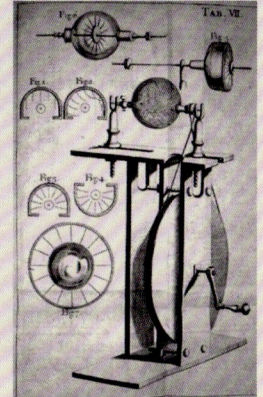

Spark of invention

As well as carrying out the experiments with vacuums, von Guericke also pursued his belief that the force of gravity was electrical in nature. Von Guericke didn't refer to the phenomena he investigated as electrical—he called them "incorporeal potencies." In 1663, he made an "electric earth" model that consisted of a sphere of sulfur, about the size of a head, with a wooden rod through the middle which was placed on supports, allowing the sphere to be rotated easily. Rotating and rubbing the sphere electrified it, so that it attracted chaff, feathers, and similar small objects. Von Guericke also observed the phenomenon of electrical conduction, noting that a thread attached to the globe would show electrical attraction at its far end. He noticed that sometimes the sulfur globe glowed when it rotated. This was due to the phenomenon of electroluminescence, the conversion of electrical energy into light. It was the first time this had been seen. Von Guericke's rotating sphere was the first electrical machine. Soon it became the standard way of producing electricity, and remained so for over a century.

The impressive hemisphere experiment was witnessed by none other than the Holy Roman Emperor Ferdinand III.

38 Clocks

A CLOCK IS A TIME MACHINE—A DEVICE FOR MEASURING TIME. Humans have always been aware of time passing: day turns to night, seasons change, the Moon waxes and wanes. As civilizations grew in complexity, so the need grew to invent ways to keep track of time more accurately.

The oldest known weight-driven mechanical timekeeper was installed at Dunstable Priory in England in 1283. By 1300, similar devices were being constructed for churches and cathedrals across Europe. Because they indicated the time by striking a bell, they came to be called *clocca*, from the Latin word for bell. The motive force for the mechanical clock was a descending weight which turned a set of gear wheels. But what was truly revolutionary was the escapement. This controlled the rotation of the gear wheels and transmitted power to the oscillator, which regulated the speed at which the clock operated. The name of the insightful inventor of the escapement is lost to time.

Galileo Galilei was one of the first to experiment with pendulums as timekeepers. His discovery that the period of a pendulum's swing was independent of the arc of its swing was crucial, but it was Dutch astronomer and mathematician Christiaan Huygens who devised the first pendulum clock in 1656.

Huygens observed that any variation in the extent of the pendulum's swing would cause the clock to gain or lose time. He devised a suspension that caused the pendulum bob to move in a cycloid-shaped arc rather than a circular one. This meant that it would oscillate in the same time regardless of its amplitude. Pendulum clocks were about 100 times more accurate than their mechanical predecessors. In 1675, Huygens made further improvements when he devised the spiral balance spring, which regulated the

A technical drawing of Christiaan Huygens' pendulum clock from the 1660s.

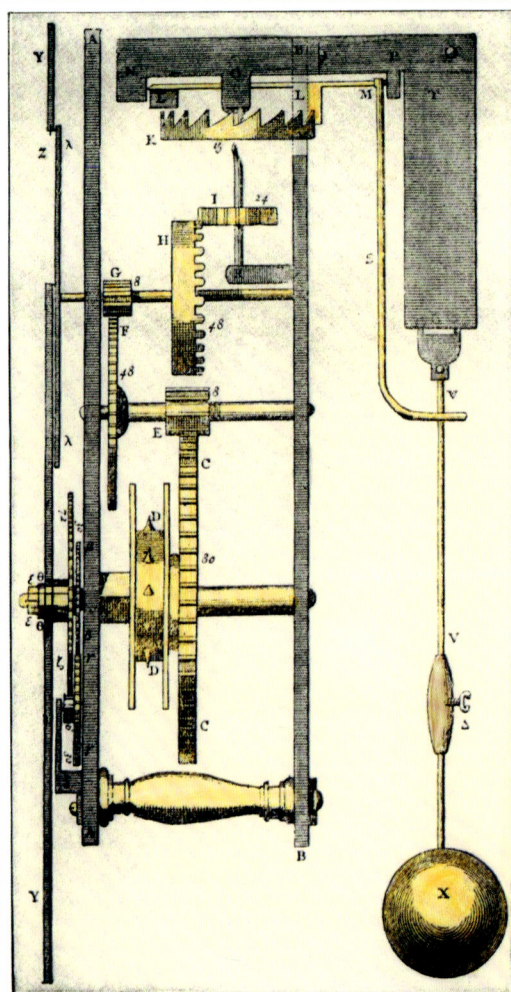

ATOMIC CLOCKS

Every chemical element and compound absorbs and emits electromagnetic radiation at characteristic frequencies. These constitute a potential atomic "pendulum" that can form the basis for very accurate clocks. In 1949, the first atomic clock was built, based on resonances in the ammonia molecule, but it failed to perform well. Cesium was more promising, and the first cesium atomic frequency standard was built at the National Physical Laboratory in England in 1955 (above). In 1967, the cesium atom's natural frequency was formally recognized as the new international unit of time, with a second defined as exactly 9,192,631,770 oscillations, or cycles, of the cesium atom's resonant frequency.

motion of a balance wheel. This rotated one way and then the other in a repeating cycle.

The first quartz clock was built by Warren Marrison and Joseph Horton in 1927. In these devices, an electric current causes a quartz crystal to resonate at a frequency that is far higher than a pendulum's oscillations. Because quartz clocks rely on a mechanical vibration whose frequency depends on the size, shape, and temperature of the crystal, no two can have the same frequency and they have to be calibrated against a standard.

CHINESE WATER CLOCKS

The water clock, or clepsydra, invented about 1500 BCE, relied on the steady flow of water to mark time. The water clock was more reliable than burning candles or oil lamps, although variations in pressure affected the water flow. Late Chinese clepsydras, developed from 200 to 1300 CE, used a water-driven bucket system to power mechanisms used to illustrate astronomical phenomena, such as rotating celestial globes.

39 Thermometers

IN 1596, GALILEO GALILEI INVENTED AN INSTRUMENT THAT INDICATED TEMPERATURE DIFFERENCES BUT COULD NOT MEASURE THEM. This predecessor to the thermometer, the thermoscope, could show if temperatures were higher, lower, or the same, but was unable to quantify the difference.

Italian Santorio Santorio is generally credited as the inventor of the thermometer around 1612, as he was the first to apply a scale to an air thermoscope. Santorio's instrument was not very accurate, since the effects of changes in air pressure on the thermometer were not understood at that time. Toward the end of the 1600s, French physicist Guillaume Amontons developed a means of measuring temperature based on a proportional change in pressure, a relationship known as Amontons' law.

Ole Rømer, a Danish astronomer, pioneered the use of mercury in thermometers, something that was improved upon by Gabriel Fahrenheit, who was the first to offer accurately calibrated devices, in 1714. Mercury expands evenly and predictably, giving accurate instruments. Fahrenheit chose 32 as the freezing point as this produced a scale that would not fall below zero under any circumstances he could reproduce. The Fahrenheit scale is still in use today, although most people use the 0 to 100 Celsius scale, which is based on the freezing and boiling points of water. In 1867, Thomas Allbutt, an English doctor, invented the medical thermometer. It was 15 cm (6 in) long and took five minutes to take a person's temperature.

An original Fahrenheit thermometer is still in use to show the temperature in a Polish fish market.

40 External Combustion Engines

AN ENGINE IS A DEVICE THAT TURNS HEAT ENERGY INTO USEFUL MOTION. The heat is delivered by burning, or combusting, a fuel, and the first successful engine systems burned the fuel outside, or external to, the moving parts.

The title of "first engine" is often given to the aeolipile, the steam-powered turbine built in the 1st century CE by Heron of Alexandria (we even do it in this book!). However, the aeolipile is not quite worthy of the full definition of engine. Heron never put its motion to use, nor did its design lend itself to that purpose. Nevertheless, the aeolipile's heat source was external to the boiler that turned water into steam, which then drove the mechanism into motion. The same was true of all other steam engine designs which appeared at the start of the Industrial Revolution in the 1700s.

MINER'S FRIEND

In 1698, Thomas Savery patented a machine—which he described as "The Miner's Friend; or, An Engine to Raise Water by Fire"—that used steam pressure to pump water from flooded mines. Savery's machine consisted of a boiler above a brick furnace, a water-filled reservoir, and a series of valves. Steam was fed into the reservoir, pushing the water it contained out through a valve until the reservoir was empty. Coolant water was then sprayed on the reservoir, causing the steam inside to condense, thus creating a vacuum that drew up more water through a second valve. Slowly and not that surely—the cast iron components often split—water could be raised from a flooded mine, although there was a limit to how high the machine could lift water.

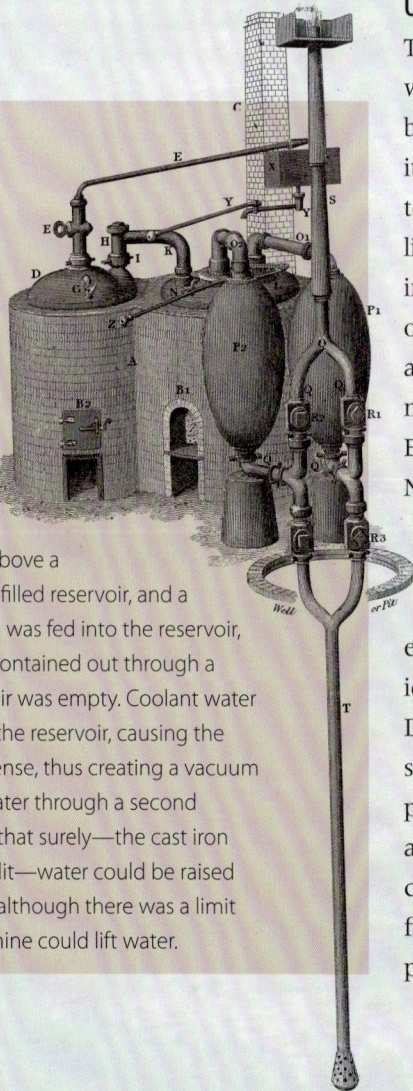

Useful work

The first steam engine that performed a useful function was the Miner's Friend, a pump system built in England by Thomas Savery (see box, left). However, the energy it harnessed could be put to work in only a few, limited ways. The crucial innovation that created, or at least would lead to, a general-use engine was made in 1712 by another English engineer, Thomas Newcomen, who developed a system using a redesigned steam engine.

Newcomen's atmospheric engine was inspired by the ideas of French scientist Denis Papin, who had suggested using high-pressure steam to push on a piston inside a gas-tight cylinder. As the cylinder filled with steam, the gas pushed up the piston,

Claude-François-Dorothée, Marquis de Jouffroy d'Abbans, invented a steam-powered boat in 1783, making a trial run on the Saône River in France.

TREVITHICKS,
PORTABLE STEAM ENGINE.

Catch me who can.

Mechanical Power Subduing
Animal Speed.

THE LOCOMOTIVE OF 1808.
From Harper's ' A Cornish Giant.'

FIG. 5.

In 1803, British engineer Richard Trevithick built the first railroad steam locomotive. By 1850, Britain had nearly 10,140 kilometers (6,300 miles) of railroad track.

which raised a levered connecting rod. Once the cylinder was full, a spray of coolant water was used to condense the steam inside back into water, which drained out. That created a vacuum, so atmospheric pressure pushing down from above forced the piston back into the cylinder, and the connecting rod was moved down.

Up, down, and round and round

Repeating this cycle of up and down strokes created a corresponding motion in the connecting rods. Newcomen's engine was first used to power pumps that drained coal and ore mines. The rod could be connected to drive a wheel, but the atmospheric engine was too weak and inefficient to make this practical. To provide any useful work, the Newcomen engines had to be the size of a house.

The name James Watt is much more closely related to steam engines than the designers who came before him. This is because, in the 1760s, the Scottish engineer made modifications to Newcomen's engine. The first was to add a separate condensing chamber so there was no need to cool the cylinder and then heat it again. Keeping the cylinder and piston at steam temperature while the engine was operating brought fuel costs down by about 75 percent. In addition, Watt boosted the power by applying steam to both sides of the piston. Watt's breakthrough engine also had an arrangement of gears that converted the up-and-down movement of a piston into circular motion. A heavy flywheel which kept turning under its own inertia evened out any jerky variations in the force delivered to the engine shaft by the action of the piston in the cylinder, and a governor connected to the flywheel regulated the flow of steam to the engine's cylinder.

Watt's engines were used to operate mills and machines. (Watt invented the unit of horsepower to promote the power of his engines, and today the watt [W] is the international unit of power.) By 1800 Watt and his business partner Matthew Boulton had installed 500 engines.

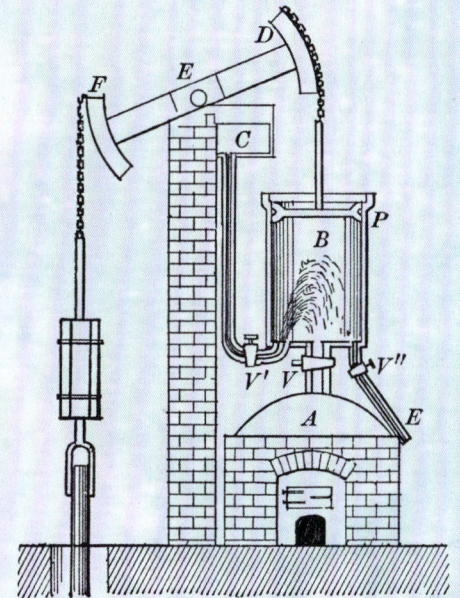

Newcomen's atmospheric engine produced an up-and-down reciprocating motion for powering pumps.

THE STEAM MAN

The potential of engine technology was obvious from the earliest days; Heron had imagined a world where all the work was done by steam-powered laborers. In 1868, the American Zadoc Dederick patented a design for a steam locomotive that walked instead of rolled. His Steam Man of Newark (above) was meant to be for sale at $300 each, but Dederick only made one prototype.

41 Spinning and Weaving

SPINNING AND WEAVING CLOTH IS AN ANCIENT SKILL. A model loom was found in an Egyptian tomb from 4,000 years ago, yet the process was not truly revolutionized by technology until the Industrial Revolution.

A loom works by interweaving strands of yarn to form cloth—the warp threads lie lengthwise along the loom and the weft threads are woven between them using a shuttle. For millennia this was done by hand and was a slow process until John Kay invented the flying shuttle in 1733. Pulling a cord sent the wheeled flying shuttle moving on a track through the warp. It greatly sped things up, and could also be used to weave wider fabrics. The success of the flying shuttle increased the demand for yarn and attention turned to making spinners more productive. Lewis Paul and John Wyatt patented a donkey-powered spinning machine in 1738 that worked by drawing fibers through pairs of spinning rollers.

In 1764, James Hargreaves invented the spinning jenny—an eight-spindled spinning wheel that could be operated by just one worker. Aggrieved spinners broke into Hargreaves' home and destroyed his machines. Richard Arkwright's water frame from 1769 used the power of a waterwheel to produce high quality thread much faster than hand-turned machines, and in 1771 he built the first textile factory to capitalize on his invention. Ten years later, Samuel Crompton invented the spinning mule, which was even more efficient than the jenny.

Sewing is the final act of making clothes from cloth. The diagram above was submitted by Thomas Saint for his sewing machine, patented in 1790.

SPINNING WHEELS

Believed to have been invented in India around 700 CE, the spinning wheel sped up the process of producing yarn by mounting a spindle horizontally and turning it by means of a large wheel rotated by hand while the operator fed in the fibers to be spun. The spinning wheel spread to Europe in the medieval period and was the mainstay of the textile industry until the mechanization of the Industrial Revolution.

Adding power

Edmund Cartwright patented a powered loom in 1785. The powered loom mechanized the process of weaving and produced cloth much faster than a hand-loom weaver could hope to achieve. As the Industrial Revolution progressed, textile factories sprang up as the spinning mules and powered looms were converted from water to steam power. By the middle of the 19th century, nearly 300,000 power looms were in use in Britain.

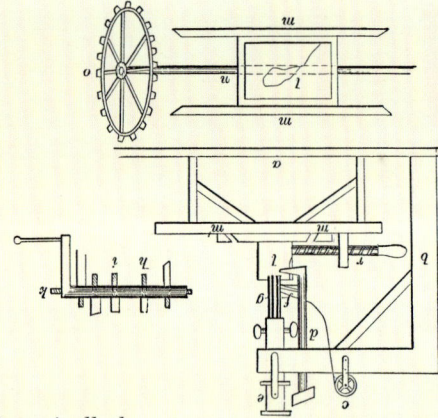

A spinning jenny in action circa 1880. (A jenny is a female donkey or mule.)

42 Ice and Refrigeration

CONTROLLING FIRE AND HEAT IS ONE OF THE OLDEST TECHNOLOGIES, DATING BACK PERHAPS A MILLION YEARS. The process for doing the opposite—making things cold—was only discovered in 1748.

A Siebe and Harrison ice-making machine from 1862. This machine was introduced by James Harrison in Australia in 1859, and the design was then improved by Augustus Siebe of London in 1861. It operated by using ether as the refrigerant; the liquid's sudden expansion into a gas was enough to freeze water. In those days, the purpose of this technology was mostly to create ice, which was then delivered to homes for use in ice boxes.

A kitchen fridge keeps food cold by expanding a gas very rapidly. This process draws heat from the surroundings, reducing the temperature. The fridge then compresses the gas back into a liquid (the whirring that clicks on and off is the pump doing this work) and the process begins again. This phenomenon was discovered in 1748 by William Cullen, one of the world's first chemistry professors. He and his students were experimenting with thermometers and found that volatile chemicals appeared to bring the temperature down as they evaporated away. Cullen set up a low-pressure apparatus where a volatile spirit would evaporate at great speed. When switched on, the device rapidly iced up.

About 50 years later, Oliver Evans, an American, proposed the compression and expansion cycle to create a mechanical refrigerator, however, like Cullen before him, he did not pursue this innovation. In 1830, Jacob Perkins, once an assistant to Evans, stepped up to build the first fridge—actually an ice machine. Continuously refined, mechanical refrigerators were at first massive industrial devices, but by the 1920s, the mechanisms were successfully miniaturized for use in the home.

THE ICE PIT

Before mechanical refrigeration, the only way to keep things cool was by making ice. Ice tech reached its peak in Persia 2,500 years ago. Ice was made on winter nights in shallow pools that were shaded from sunlight by a wall. It was then stored in a *yakhchal*, or domed ice pit (below), which kept it frozen through the summer. Warm air rose inside the dome and whistled out through vents in the top. The outside wall of the yakhchal was covered by straw during the day to protect it from sunshine. At night the straw was removed to allow heat from the bricks to radiate away.

43 Vaccination

ENGLISH COUNTRY DOCTOR EDWARD JENNER WAS INTRIGUED BY HOW
milkmaids who had contracted cowpox from cows were immune to the
more deadly smallpox. Through this, he found a way to prevent disease.

*Jenner's young patient
is not pleased with the
treatment he is getting.*

A rudimentary form of smallpox inoculation was developed in China
around 900 CE. This included variolation, or the introduction of material
from smallpox pustules into the skin of a healthy patient. It was practiced in
Europe and North America in the 18th century, but it could disfigure the skin
or kill, so few took the risk.

In May 1796, Jenner tested out his ideas concerning the protective
properties of cowpox on eight-year-old James Phipps, rubbing cowpox
material into scratches on the boy's arm. Within a few days the boy became
mildly ill with cowpox, but soon recovered. Jenner next inoculated the
boy with material from a smallpox lesion. To everyone's great relief, Phipps
did not develop smallpox, either then or in the future. Jenner was correct:
cowpox did indeed protect against smallpox. The technique of introducing
material under the skin to confer protection against disease became
universally known as vaccination, from the Latin name for the cow, *vacca*.

44 Batteries

**BEFORE THE 18TH CENTURY, ELECTRICITY COULD NOT BE
STORED OR PUT TO MUCH USE.** The inventions that solved
that problem still form the basis of the modern world.

Around 1780, Italian physicist Luigi Galvani discovered that he
could make a dead frog's legs kick by applying an electric spark to
the frog's spinal cord. He was particularly interested in the fact that
muscle contractions also occurred when he used his iron scalpel to
dissect the leg of a frog fixed to a copper hook. Galvani decided that
there must be a type of electrical fluid in the frog's body, naming it
animal electricity. In public demonstrations, Galvani would show
rapt audiences dozens of frogs' legs twitching away on copper
hooks hung on an iron wire.

Fellow Italian Alessandro Volta was intrigued by Galvani's
discovery, but he didn't believe the frogs were producing electricity.
He suspected that it was contact between the different metals that

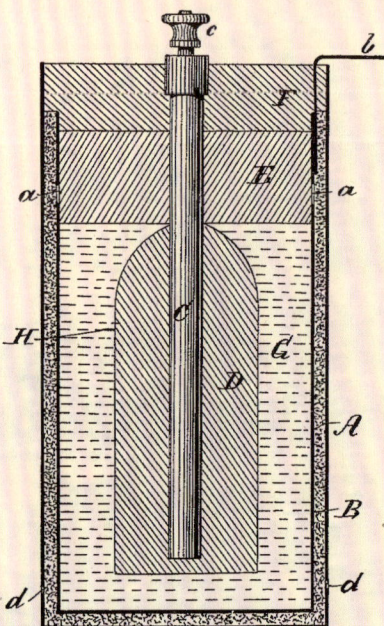

THE LEYDEN JAR

In 1745, Dutch physicist Pieter van Musschenbroek invented a device for storing electricity. Named after the town in which he lived, the Leyden jar was a glass jar filled with water, with metal foil lining the inside and around the outside. Curious to see if electricity was indeed being stored, van Musschenbroek touched the inside and outside of the jar simultaneously. He thought "it was all up with me" as the charge coursed through him. Today, we might think of the Leyden jar as an early form of capacitor. Several jars connected in parallel increased the amount of charge stored. Van Musschenbroek steadfastly refused ever to repeat his shocking experience again.

was generating the electricity. There being no suitable detector available to test his hypothesis, Volta relied on popping various combinations of metal into his mouth to see what happened.

Deciding zinc and copper gave the best results, Volta built a vertical pile of alternating zinc and copper discs, separated by circles of cloth soaked in brine. When he connected a wire to each end of the pile and touched it with his fingers he felt "a small prickling or slight shock." He had created a flow of electricity. The more disks he added to the pile, the greater the shock he felt.

For the first time, Volta had made it possible to produce a steady supply of electrical power. Volta reported his invention to the Royal Society in London in 1800, and within weeks the voltaic pile, as it came to be known, was being put to use in laboratories. Scottish chemist William Cruickshank designed a horizontal version of the pile, with the zinc and copper plates housed in an insulated wooden box, that became very popular. The experiments Volta's pile made possible were soon firing the imaginations of scientists across the world.

Left: Volta's pile. The term "pile" is still applied in some languages. The English term "battery" was coined by Benjamin Franklin, who thought that a bank of connected Leyden jars resembled a battery of cannons.

Far left: Today's batteries use the dry cell design, as in this early 1890 version.

Right: Today's compact and high-powered rechargeable batteries are based on lithium-ion technology.

RECHARGEABLE BATTERIES

The rechargeable battery has become an almost essential part of modern life. Its ancestor, the lead-acid cell, still the most common vehicle battery today, was invented by Gaston Planté in 1859. It consisted of coils of lead and lead oxide separated by rubber strips and held in a bath of dilute sulfuric acid, which reacted with the lead to produce a current. When current was added to the battery from an outside source, the process reversed, and the battery was recharged.

45 Electric Light

THE STORY OF THE ELECTRIC LIGHT IS NOT ALWAYS CLEAR, LET ALONE ILLUMINATING. Many people were involved in inventing the device, the technology of which continues to evolve even today.

For centuries, light was produced by burning oil in lamps or wax in candles. That changed in 1809, when the British scientist Sir Humphry Davy used 2,000 electric batteries to produce the brilliant light of an arc lamp. Davy, director of the laboratory at the Royal Institution in London, passed an electric current to two charcoal rods, causing a spark to jump across the gap and complete the circuit. The carbon became so hot that its ends became incandescent, glowing with a brilliant light. If anything, the light was too bright to be used in domestic situations, so it was mainly limited to street lighting. In addition, before electricity was widely available, it was difficult to power the lights on a large scale.

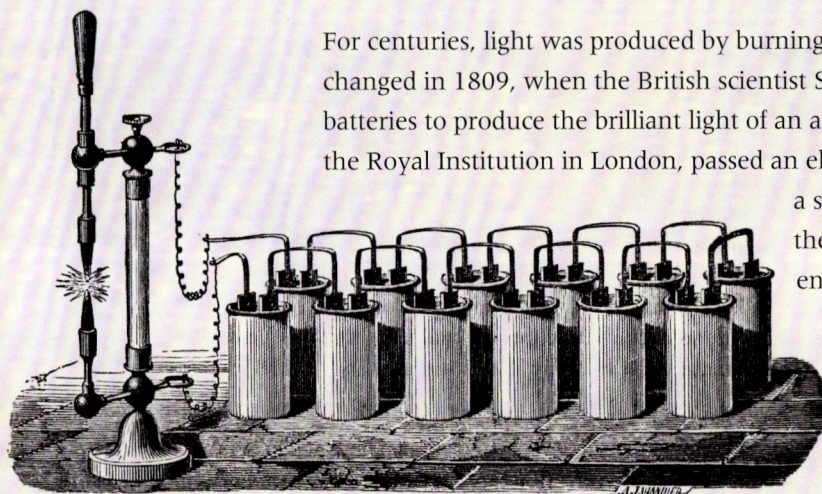

An arc lamp of the kind pioneered by Humphry Davy. Ever the showman, he used them in public demonstrations, but they proved unsuitable (not least due to their incredible inefficiency) as a means of lighting.

Incandescent light

The next step on the way to electric light was also a dead end. In 1835, an obscure Scottish inventor named James Bowman Lindsay demonstrated a light bulb he had invented. Electricity was more readily available around now to power bulbs that depended on heating a thin metal filament until it glowed. In order to prevent the heat reacting with oxygen in the air and bursting into flame, the filament was contained within a glass bulb from which most gas had been removed.

For reasons that are not clear, Bowman seems to have lost interest in his invention. There would be no practical incandescent light bulb for another 40 years, when two inventors came up with their own versions at nearly the same time. The British inventor Joseph Swan created a light bulb in 1878, and Thomas Edison, inventor of the phonograph, came up with his own version just a year later.

GEISSLER TUBES

In 1857, the German physicist Heinrich Geissler began to create long, sealed glass tubes filled with gases, fluids, or minerals that reacted to electricity. When Geissler applied an electric current through either end of the tube, the splitting and recombining of electrons and ions caused the gas to fluoresce, or glow, often with brilliant colors. These popular novelties were the first gas-discharge lamps and led to neon lighting and modern low-energy bulbs.

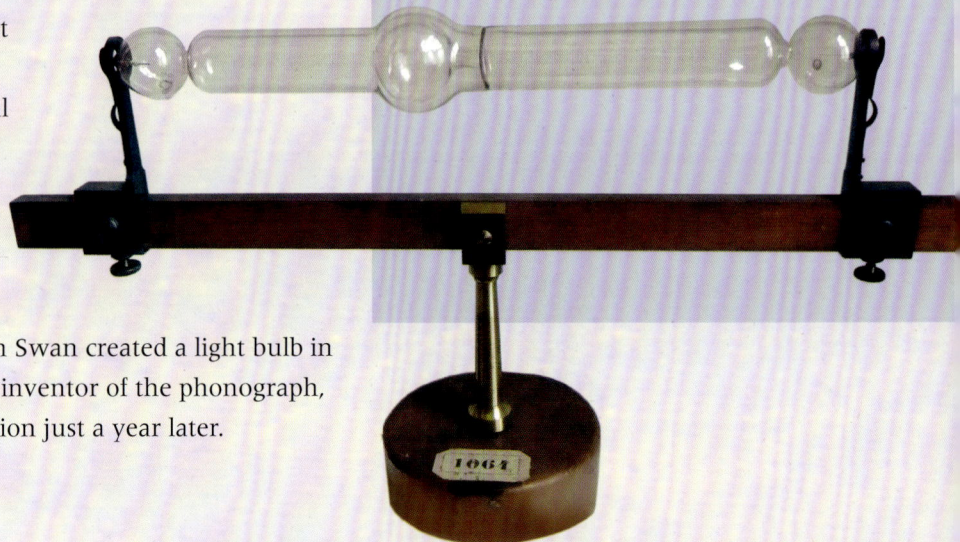

LOW-ENERGY BULBS

Incandescent bulbs containing a heated filament use a lot of energy. Since the 1970s, they have been widely replaced by low-energy bulbs. Some of these are compact fluorescent bulbs (CFBs). These are coiled and folded gas-discharge lamps that send an electrical current through a tube of mercury vapor. The vapor gives out invisible ultraviolet light which is converted to a visible glow by the inner lining of the tube. Others are light-emitting diodes (LEDs), which use semiconductors that light up when electrified.

Edison's filament was a piece of carbonized bamboo, a material he chose after hundreds of tests because it was both cheap and long-lasting. While Edison and Swan fought over taking credit for the new device—they eventually agreed to go into business together—Edison continued his work on developing a reliable electricity supply to homes to encourage uptake of the new device.

Energy savings

The light bulb transformed the world. It extended the length of the day, as well as making it safer to be outdoors or to travel. Variations on light bulbs continued to appear, from powerful searchlights to brightly colored tubes holding colorfully glowing, electrified gas—so-called neon lights. Today, incandescent light bulbs, which produce more heat than light, are being replaced by energy-saving bulbs, a process that is sped up in some countries by government legislation.

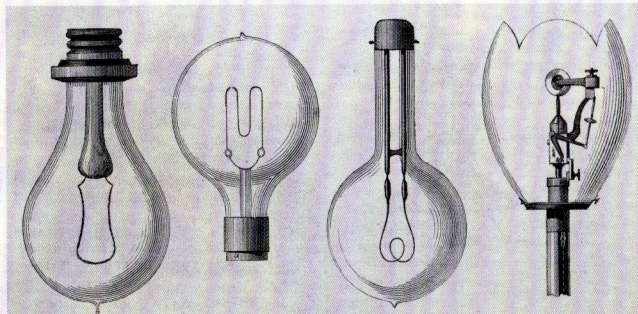

Four early designs of light bulb, as seen at the 1881 International Exposition of Electricity in Paris. From left to right, bulbs invented by Edison, Hiram Maxim, Swan, and the Werdermann lamp system.

46 Anesthetics

TODAY, ANESTHETICS ARE ROUTINELY ADMINISTERED SO THAT SURGERY CAN BE CARRIED OUT SAFELY AND PAINLESSLY. Before their discovery surgery was such an excruciating and terrifying ordeal a patient might opt for death instead.

According to Chinese tradition, physician Hua Tuo developed a herbal potion called *mafeisan* in the 2nd century CE. He is said to have administered it to a patient with an arrow wound who reportedly played chess unconcernedly while his bone was scraped clean. Japanese surgeon Hanaoka Seishū attempted to replicate Hua Tuo's formula toward the end of the 18th century, with some success, carrying out the first operation under general anesthesia in 1804. The anesthetic powers of various gases were explored in the 18th and 19th centuries. In 1799, Humphry Davy described the pain-relieving effects of inhaling nitrous oxide and suggested using it to ease pain during operations. Michael Faraday was one of the first to study the inhalation of ether, publishing his findings of its soporific effects in 1818, and in 1846 at the Massachusetts General Hospital in Boston, William Morton gave the first demonstration of its value for operations.

Hanaoka Seishū administered his herbal anesthetic, which used a number of toxic plants like wolfsbane and jimson weed, with a soaked cloth held on the face.

47 Internal Combustion Engines

SMALLER, MORE EFFICIENT, AND MUCH MORE POWERFUL THAN COAL-POWERED STEAM ENGINES, petroleum-fueled engines made travel across oceans and on roads widely available, but have taken their toll on the environment.

At first glance, a blueprint of an internal combustion engine does not look entirely different to the external system, with motion being transmitted from pistons moving up and down inside cylinders. The big difference is the source of gas that pushes the pistons. In the internal combustion engine the gas is the exhaust product from a controlled explosion of fuel. The first internal combustion engines were intended to be stationary power sources. However, with no need for a furnace and boiler, the internal combustion engine is much more compact and lightweight than a steam engine, and ideal for use in vehicles (it was also the power source of early aircraft).

FIG. 7.—Lenoir's Gas Engine.

A 3-horsepower stationary internal combustion engine constructed by Étienne Lenoir in 1861.

THE STIRLING ENGINE

Named for its inventor, a Scottish priest called Robert Stirling, and hailing from 1816, the Stirling engine is therefore almost as old as the internal combustion engine. The Stirling engine (below) works by two pistons moving in a single cylinder. The cylinder is sealed, contains a fixed amount of gas, and has a hot end and cold end. The cyclical differences in pressure move the pistons. The hot end is warmed by a fuel burning outside the cylinder, so this is actually an external combustion system.

Inside history

An 1807 design by Frenchman François Isaac de Rivaz used a mixture of hydrogen and oxygen gas as a fuel, which was ignited by a spark to make steam. Throughout much of the 19th century, engineers competed to develop more efficient internal combustion engines. In 1860, the Belgian Étienne Lenoir built the first single cylinder, two-stroke engine. In other words the ignited fuel-air mixture forced the piston into a down stroke,

DIESEL ENGINES

In 1895, Rudolf Diesel invented an engine that could use heavier, thicker oil than gasoline. The oil he used is harder to ignite, but burns hotter. Instead of using a spark to ignite the fuel, now called diesel, the piston compresses the vapor so much that it heats up and explodes. Diesel engines (the first one is shown left) need to be large and robust to do this, and are best suited to power trucks, trains, and ships.

driving a wheel. The up stroke then trapped fuel vapors ready for the next ignition. In 1864, the German pair Nicolaus Otto and Eugen Langen built a four-stroke (see diagram, below) internal combustion engine, more efficient than a two-stroke, and in 1876 this was refined by compressing the fuel mix prior to combustion to achieve far higher efficiency. This is the cycle, named for Otto, that is used in gasoline-powered autos today.

In 1885, another German, Gottlieb Daimler, managed to create an internal combustion engine small enough to power the Reitwagen, a two-wheeled vehicle that was the world's first motorbike. The same year Karl Benz added a third wheel to make the first car (see more, page 71).

1. The piston moves down inside the cylinder, drawing an air-fuel mixture into the vacuum created. 2. As the rotating crankshaft pushes the piston up again, the air-fuel mix is compressed. 3. A spark plug ignites the fuel. The explosion pushes the piston down and the crankshaft around. 4. The piston is pushed up, forcing the exhaust gases out.

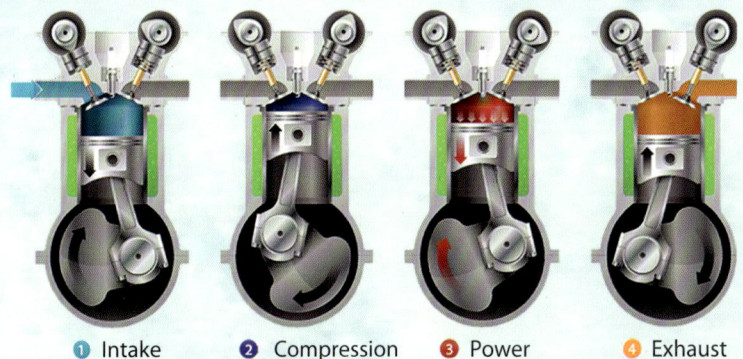

❶ Intake   ❷ Compression   ❸ Power   ❹ Exhaust

48 Canning

SINCE THE EARLIEST TIMES PEOPLE HAVE LOOKED FOR WAYS TO PRESERVE THEIR FOOD SUPPLIES. Canned foods are commonplace today, but are a fairly recent development in the long history of food preservation.

In 1795, Napoleon Bonaparte offered a reward for whoever could develop a safe, reliable food preservation method for the Grande Armée of France. Fifteen years later, in 1810, Nicolas Appert developed a method of heat-processing food and sealing it in reinforced glass jars sealed with wax.

Around the same time, Peter Durand in England preserved food in unbreakable iron cannisters lined with tin. This was the first true "canning" (as opposed to "bottling") method. It was an expensive process, as each can was sealed by hand and cooked for six hours. (And as the can opener had yet to be invented, getting at the food was problematical.) In 1846, Henry Evans invented a die-cast process in England that increased can production tenfold, and the following year Allen Taylor in the United States patented a machine-stamped tin can. The first commercial cannery was built in 1912 by Thomas Kensett. The canned food market would never look back.

The contents of this can of tomatoes from 1920s Spain are probably perfectly good to eat.

49 Photography

THE CAMERA OBSCURA, FIRST DESCRIBED IN THE 10TH CENTURY BY AL-HASAN IBN AL-HAYTHAM, is a space with a small hole in one side. Light entering through the hole casts an inverted image of the scene outside onto the opposite wall. This simple setup was the ancestor of photography.

It was one thing to form an image, but how could it be captured? In 1727, German chemist Johann Schulze accidentally discovered that silver salts darkened when exposed to sunlight. It took another hundred years before Schulze's salts and the camera obscura were brought together to take the first photograph. Frenchman Nicéphore Niépce devised a method of using light to copy engravings. He oiled an engraving to make a transparent print, placed it on a light-sensitive bitumen, and exposed it to sunlight. The solution under the light areas hardened, while that under the dark areas stayed soft and could be washed away, leaving an exact copy of the engraving. Niépce called his process, perfected in 1822, heliography. In 1826, Niépce used heliography to produce the first successful photograph on a pewter plate.

The first photo was of the view of the courtyard of Niépce's country estate as seen from an upstairs window of the house. The exposure time was about eight hours.

Theatrical scene painter Louis Daguerre often traced images produced by a camera obscura for his sketches. In partnership with Niépce, he searched for better ways of capturing the image. In 1835, three years after Niépce's death, Daguerre discovered that an image formed on a plate coated with silver iodide salts, and that it could be made visible by exposing it to mercury vapor. The great advantage of this method was in reducing exposure times to 30 minutes. However, the image gradually darkened until Daguerre found a way of using a solution of salt to dissolve the unused silver iodide. Using this process, he produced a photograph, called a daguerreotype.

In 1835, English scientist William Henry Fox Talbot made light-sensitive paper by soaking it in solutions of common salt and silver nitrate to produce silver chloride. When exposed to light, the silver chloride formed a "negative" image, where light objects are dark and vice versa. Positive images could be made from this by simply bringing more paper into contact with the negative and exposing it to light. The positives were then "fixed" by washing away the unexposed chemicals.

DIGITAL IMAGES

Photography in the 21st century has been transformed from a reliance on chemically developed light-sensitive emulsions to digital technologies that allow images to be captured and stored, and manipulated, in ways previously unimagined. The first consumer digital cameras appeared in the late 1980s. Today, anyone with a smartphone can capture images that would be the envy of previous generations of photographers.

50 The Motor

IF AN ELECTRIC MOTOR IS DEFINED AS A DEVICE THAT CONVERTS ELECTRICAL ENERGY INTO KINETIC ENERGY, then English physicist Michael Faraday invented the first electric motor on Christmas Day, 1821.

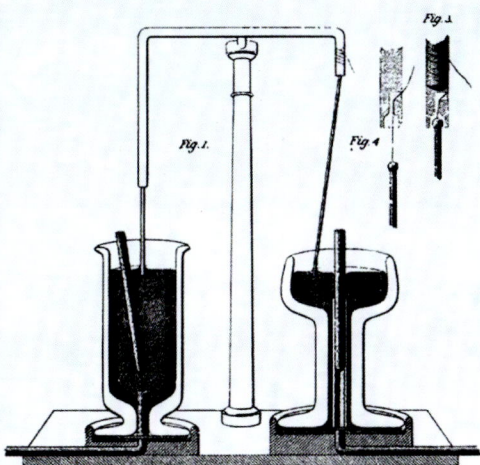

Michael Faraday's rotating contraption showed the principle of electric motors but was unable to put the motion to good use.

In 1820, Danish physicist Hans Christian Ørsted discovered the connection between magnetism and electricity. This fundamental link—that electric conductors become magnetic—makes possible all of the electricity generated in the world's power stations and the countless electric motors we rely on.

A year later, Faraday set up an ingenious experiment in which he prepared a beaker of mercury with a bar magnet fixed vertically at the center. A freely moving wire dipped into the mercury, completing an electrical circuit. When Faraday connected his experiment the current flowing through the wire produced a magnetic field in the wire that interacted with the field from the bar magnet, causing the wire to rotate.

In 1822, Peter Barlow replaced the dangling wire with a spoked wheel that dipped into a trough of mercury, so a current began to rotate the wheel. Prussian physicist and engineer Moritz Hermann von Jacobi built the first practical electric motor in 1834. His electric ferry boat, powered by battery cells, could carry 14 people across a river.

Around the same time, Thomas Davenport, a Vermont blacksmith, devised an apparatus that used electromagnets mounted on a rotor and on fixed poles. When he switched on the current, the interaction between the magnets caused the rotor to make a half-turn. Davenport discovered that if he reversed the wires to one of the magnets the rotor would complete another half-turn. He then devised what is now called a brush and commutator. As the wheel turns, its motion breaks the circuit and instantly reconnects with an opposite polarity. That reverses the polarity of the electromagnets, pushing the wheel through the next half-turn. As the process is repeated, so the wheel spins. This is the system that is still used in electric motors today.

ELECTROMAGNETS

In the 1820s, André-Marie Ampère placed an iron rod inside a coil of wire and found that it behaved like a magnet when the current was turned on. In 1824, William Sturgeon constructed the first U-shaped electromagnet by winding a few turns of copper wire around an iron horseshoe. Five years later, Joseph Henry improved the idea by using many turns of insulated copper wire around a thicker, U-shaped iron core, eventually building a magnet that could lift a tonne of iron.

51 Computers

TODAY, WE TAKE FOR GRANTED OUR ACCESS TO VAST AMOUNTS OF DATA. We have more computing power in our pockets than was available to an entire nation less than a hundred years ago. Perhaps surprisingly, the roots of computing lead back to silk weaving

Only a small part of the Analytical Engine was ever built. This trial version could handle 25-digit numbers represented on dials. Babbage's original vision was to handle 10,000 numbers with 40 digits each.

In 1801, Joseph-Marie Jacquard invented a way of automating the highly repetitive process of weaving silk. He did this by means of a set of wooden cards punched with holes that determined which cords of the fabric warp should be raised for each pass of the shuttle. Mill owners happily embraced Jacquard's invention, but loom operators saw their livelihood threatened, and angry mobs smashed Jacquard looms.

In 1839, Francois Michel-Marie Carquillat used a Jacquard loom, employing 24,000 cards, each with 1,000 hole positions, to weave a portrait in silk of Jacquard himself. The incredible detail of the result inspired English mathematician Charles Babbage in the design of his Analytical Engine—the first general purpose programmable computer—as he saw the potential of punched cards for programming, input, output, and storage of information.

Mechanical calculator

Earlier, in 1822, Babbage had an idea for a room-sized, steam-driven calculating machine he called the Difference Engine. He obtained government funding for his project as it was hoped that Babbage's machine would eliminate errors in navigation tables. But, after 10 years of development, the complex machine was still not finished. Undeterred, Babbage became more ambitious. The Analytical Engine would be as big as a house and require not one, but six steam engines to operate it.

Babbage made an intellectual leap regarding Jacquard's punched cards. In the Jacquard loom, a thread passes or is

SOFTWARE

Software, in a general sense, is the set of instructions a computer follows to perform a task. Jacquard's cards were like software for his loom. Credit for writing the first piece of software, then and for decades afterward written on punchcards, goes to computer scientist Tom Kilburn. Kilburn and his colleagues built 'Baby,' one of the earliest electronic computers, at the University of Manchester, England. Baby was programmed to perform mathematical calculations using Kilburn's software. First run on June 21, 1948, it took 52 minutes to correctly compute the greatest divisor of 2^{18} (262,144). The term "software" was coined by mathematician John Tukey in an article on programs for electronic calculators in a 1958 issue of *American Mathematical Monthly*.

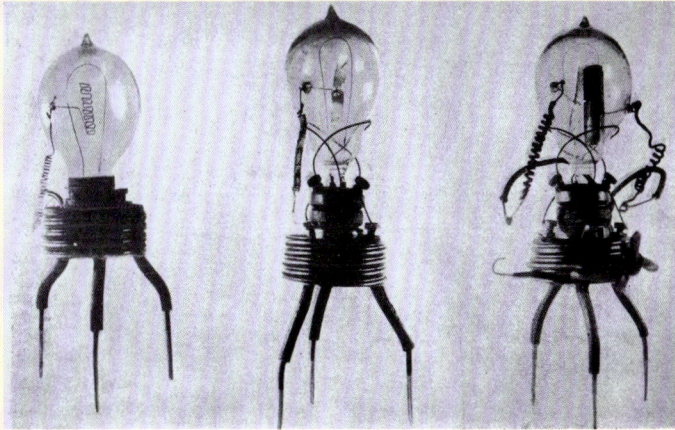

THE FLEMING VALVE

In 1904, British scientist John Ambrose Fleming made one of the most important breakthroughs in electronics. It concerned the Edison Effect, discovered by Thomas Edison, which is a flow of electrons from the hot filament of an incandescent bulb to an extra electrode. Fleming used this phenomenon to boost a weak radio signal, wiring a vacuum tube into a radio-receiving circuit. Fleming's invention became known as the Fleming valve. This was the forerunner of the diode, or electronic switch, which was used in the first electronic computers.

stopped by the presence or absence of a hole. Babbage saw that the pattern of holes could be used to represent an abstract idea such as a problem to be solved or the data required for its solution. Babbage called the two main parts of his Analytical Engine the "Store" and the "Mill." In modern terms the Store was the memory and the Mill was the processor.

Math machine

Babbage's invaluable collaborator in his endeavors was Ada Byron, daughter of poet Lord Byron and later Ada, Countess of Lovelace. Ada began writing programs for the yet-unbuilt machine, thus earning the title of the world's first computer programmer. She invented the subroutine and was the first to recognize the importance of looping, which is using an instruction that is repeated until a specified condition is obtained. She wrote, "…the Analytical Engine weaves algebraical patterns just as the Jacquard loom weaves flowers and leaves."

Babbage's Analytical Engine was never built, but the idea of programmable calculating machines with operating instructions encoded on punch cards didn't die. In 1889, American engineer Herman Hollerith, inspired by watching a bus conductor punch tickets, built a tabulating machine based on punch cards for the U.S. Census Bureau. To capitalize on his inventions, Hollerith formed a company which eventually became International Business Machines (IBM).

Z1, built by Konrad Zuse in 1937, was a mechanical computer powered by electricity. It was the first freely programmable computer in the world.

ENIAC (Electronic Numerical Integrator and Computer) was built in America in 1945, and was known at the time as the Giant Brain. It was programmed by physically wiring its 20,000 valves together.

52 Generators

IN 1831, MICHAEL FARADAY DISCOVERED THAT MOVING A MAGNET NEAR A CIRCUIT MADE AN ELECTRIC CURRENT FLOW. This process, called induction, is put to work by every electricity generator.

Nikola Tesla was famous for conducting spectacular demonstrations of electricity. This image published in 1899 shows how the eccentric inventor charging his body (while safely insulated) to make gas-discharge lamps glow.

Faraday also discovered that it was a fluctuating magnetic field that generated electric currents, and the field could be created by moving a magnet in relation to a conductor. Faraday published his results in 1832, ahead of Joseph Henry, who had also recognized the phenomenon while working with electromagnets. These discoveries led to the dynamo, or generator, which worked like a motor only in reverse, turning motion into current instead of current into motion. The Serbian-American engineer Nikola Tesla invented methods of using induction to boost the strength of a current so it can be sent long distances along power cables, thus revolutionizing the supply of electricity.

53 Electrical Communications

The original Cooke and Wheatstone telegraph console from 1837, which was soon superseded by simpler Morse code systems.

THE INVENTION OF THE ELECTRIC TELEGRAPH, WHICH SENT INFORMATION OVER great distances, was a step forward akin to the invention of the printing press 500 years earlier.

The first electric telegraph was invented in 1816 by Francis Ronalds, an English cheese merchant turned amateur scientist. His device had two clocks, the dials of which were marked with the letters of the alphabet, linked by several kilometers of insulated wire, which carried electrical impulses. Ronalds offered his invention to the Royal Navy, but was turned down on the grounds that, following the successful conclusion of the Napoleonic Wars, telegraphs were totally unnecessary!

MORSE CODE

Samuel Morse's telegraph couldn't transmit voices or written messages so, with some help from his assistant Alfred Vail, he developed a code in which characters were represented as a series of short pulses of electricity (dots) and longer ones (dashes) transformed into buzzes, beeps, or flashes of light.

After reaching across land, telegraph cables were run under the sea. North America was connected to Europe in 1858. Before that it took 10 days to send a message across the ocean by ship. It now took seconds.

Action at a distance

In 1830, Joseph Henry sent an electric current down a wire to activate an electromagnet 1.5 kilometers (.93 miles) away, causing a bell to ring. British inventors William Cooke and Charles Wheatstone used this long-distance switch system, or relay, in their telegraph system, patented in 1837, consisting of six wires activating five needles that could be made to point to specific letters and numbers according to the current sent through the wires. Cooke and Wheatstone's telegraph system was successfully introduced on Britain's railroads to transmit signals and, perhaps more importantly, time. The rapidly spreading rail network needed a timetable based on a standard time system.

Around the same time, Samuel Morse, a painter, had a shipboard conversation in 1832 concerning the recently invented electromagnet. Having little real understanding of electricity, Morse sought help from Leonard D. Gale, a colleague at the City University of New York, who was familiar with Joseph Henry's work. Gale's knowledge of Henry's experiments was vital to the development of Morse's telegraph system. He showed Morse how to boost signal strength and to use a relay system invented by Henry. Morse's sender, first demonstrated in 1837, consisted of a plate with long and short metal bars representing the dots and dashes of Morse code. The operator slid a pointer connected to a battery across the bars and the appropriate dots and dashes were transmitted. The receiver had a stylus on the end of an arm activated by an electromagnet. The stylus made a dent in a paper tape, wound past by a clockwork motor, which was read by the operator.

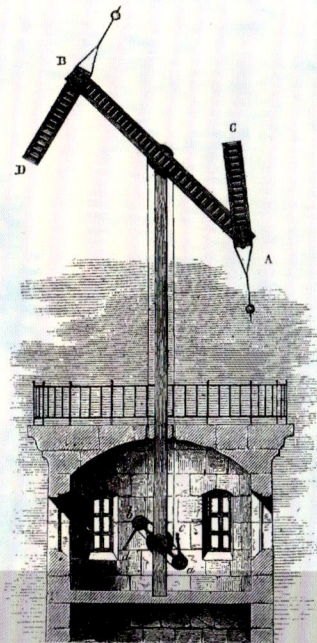

SEMAPHORE TOWERS

Before the electric telegraph, messages not actually physically carried had to be sent visually. One of the most successful such systems was the semaphore, developed in 1791 by brothers Claude and Ignace Chappe in France. Their system, which could achieve a signaling rate of three symbols a minute, consisted of pairs of movable arms, positioned to represent different numbers and letters, mounted on towers spaced 5 to 10 kilometers (3 to 6 miles) apart.

54 Fuel Cells

FUEL CELLS ARE STILL MODERN ENOUGH TO FEATURE IN FUTURISTIC DESIGNS for space colonies or sustainable homes and vehicles. However, the idea behind them—converting chemical energy into electricity—is not new.

Inside a fuel cell, hydrogen (red) is pumped into the cell, where the hydrogen nuclei are separated from their electrons (yellow). The electrons flow around a conducting loop as an electric current, which can be put to use. The hydrogen nuclei pass through a membrane (center) and then combine with the electrons and oxygen (blue) from the air to form steam, which is the only waste product.

The Welsh physicist William Grove built an experimental fuel cell in 1842, but development really began in earnest in the 1950s. All fuel cells have a positive and negative electrode (an anode and cathode) and an electrolyte, or conductive fluid, between them. Most cells are fueled by hydrogen and oxygen. Hydrogen atoms enter the anode, where they react with a catalyst so each atom loses an electron, which passes through an external circuit, forming a useful current.

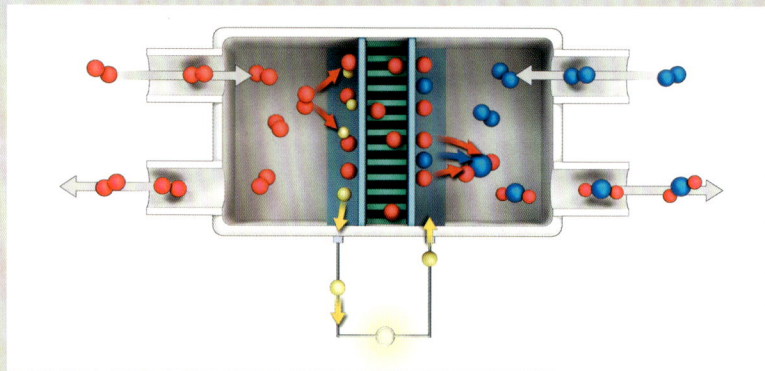

55 Elevators

AN EARLY ELEVATOR DESIGN WAS THE FLYING CHAIR, which was hoisted on ropes so that King Louis XV of France could get to his mistress's room without having to climb the stairs. However, high-rise buildings would need something safer.

While the mechanics of an elevator—a passenger car being hauled up and down by a set of strong steel cables called hoisting ropes—is relatively straightforward, its success or failure as a product rests on its safety measures. As other engineering innovations led to taller and taller buildings, in 1853 the American engineer Elisha Otis won the race to design a safe elevator. The cables from his car looped around a pulley wheel attached to a motor on the roof of the building. The weight of the car was balanced by a counterweight, which fell under gravity and thus reduced the power needed to lift the car. To prevent the elevator falling if the main cable broke, Otis built toothed guide rails on either side. Upon breaking, the cable released metal levers from the side of the car which locked into the teeth, and stopped the car from falling. In 1857, Otis's company introduced the world's first ever passenger elevator at the New York store of E. V. Haughwout. Powered by steam, Otis's elevator could transport six people at 12 m (40 ft) per minute. Modern-day elevators are capable of traveling at far greater speeds, up to 510 m (1,700 ft) per minute.

Elisha Otis demonstrates his safety system at the Exhibition of the Industry of All Nations held in New York City in 1854.

56 Dyes

FOR THE AVERAGE MID-19TH CENTURY PERSON, DECIDING WHAT TO WEAR LARGELY MEANT CHOOSING BETWEEN VARIOUS SHADES OF BROWN. Colors for garments, derived exclusively from natural ingredients, were expensive and the preserve of the rich and well-to-do. A young chemist's discovery would widen that narrow choice of colors.

In 1856, 18-year-old William Perkin was working as an assistant to August von Hofmann at the Royal College of Chemistry in London. Hofmann challenged his assistant to synthesize quinine, an expensive natural substance that was much in demand for the treatment of malaria. The process involved using coal tar, which often left a black residue in the bottom of glassware. It was while cleaning out a flask with alcohol that Perkin noticed that he had produced a purple solution. Testing the solution on silk, Perkin discovered that he had a richly colored dye that remained in the silk, even after washing. After thousands of years of costly natural dyes, made from rare plants, beetles, or even snails, the first synthetic dye that was also cheap to make had been discovered.

Perkin called the chemical mauveine, and it grew in popularity as he developed ways for it to be used on other fabrics. He borrowed money from his father to establish a factory and advised the cloth industry on how best to use the new synthetic dye. By 1860, he was rich and famous, in no small part due to the enthusiasm of Queen Victoria and other celebrities for the new dye.

Many new dyes were then discovered (some by Perkin himself), and dye factories sprang up across Europe. In the pursuit of more and more colors, the search for new dyes drove advances in chemical knowledge and even helped developments in medicine and pharmaceuticals (see box).

CHEMOTHERAPY

German biochemist Paul Ehrlich was an expert on the use of dyes as biological stains. In 1890, he began researching chemical substances that had a particular affinity for disease-causing organisms and would attack them directly while leaving the healthy cells of the body unscathed. Such chemicals would be, as Ehrlich put it, "magic bullets" fired directly at the cause of infection. Ehrlich and his colleagues tried out hundreds of drugs, before identifying arsphenamine, an arsenic-based chemical that was very effective against syphilis. Their discovery was announced in 1910. It gained Ehrlich international recognition and popularity, and he became famous as one of the main founders of chemotherapy.

It is said that people local to Perkin's lab in west London could tell which color he was researching at the time from the shade of the nearby canal into which Perkins dumped his waste water.

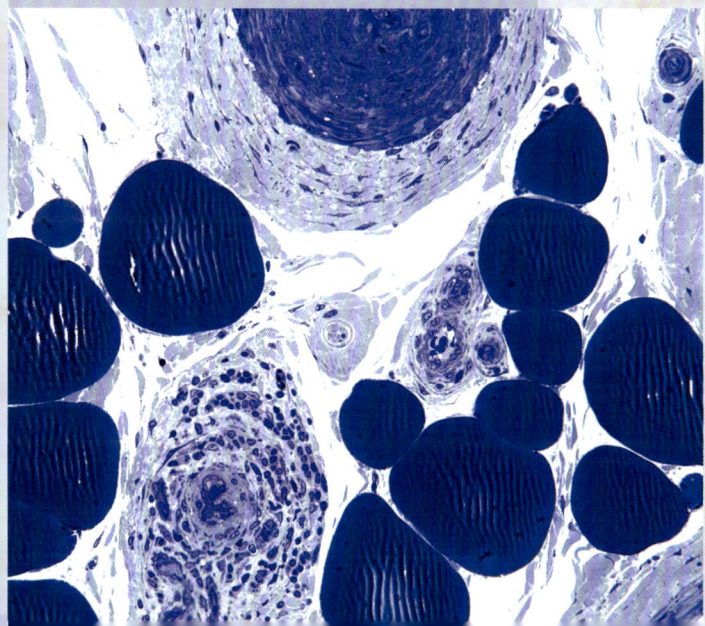

57 Plastics

PLASTICS ARE THE UBIQUITOUS MATERIAL OF THE 21ST CENTURY "AGE OF PLASTIC." From shopping bags to spacesuits, keyboards to knee replacements, plastics are inescapably everywhere.

NYLON

In 1935, Wallace Carothers succeeded in making an artificial fiber that could replace silk. This fiber, called nylon, was first used in toothbrush bristles in 1938 and during World War II it was used to make parachutes—and stockings (a revolutionary fashion item). It is a huge part of today's textile industry.

Beautiful....
ON THE LIMB

Claussner
KLEER-SHEER
Exquisite HOSIERY

For Those Who Like the Finest

Claussner Hosiery Company
INCORPORATED
PADUCAH, KENTUCKY

Plastics are strong, lightweight, and flexible, and being "plastic," they can be easily shaped and reshaped.

As the growing demands of industry for raw materials in the 19th century put pressure on natural resources, inventors looked for ways of producing artificial substitutes. In 1862, chemist Alexander Parkes exhibited his newly patented material he called Parkesine. Made from cotton fibers dissolved in nitric acid and mixed with vegetable oil, it was the first artificial plastic.

Celluloid was developed in 1869 by John Wesley Hyatt. Experimenting with pyroxylin and camphor, he patented a process in which the two ingredients were pulverized then combined into a single mass, which was drained of water under pressure, and molded at high temperature. The resulting substance was light, strong, and versatile—but also highly flammable. One of its most important uses was as part of the film that became the basis of the movie industry, before celluloid was phased out in the 1950s.

American chemist Leo Baekeland's Bakelite was the first thermoset material, meaning it was moldable on first heating but then kept its shape when heated again.

Chemical chains

Plastics such as Parkesine and celluloid are synthetic polymers, which are long-chain molecules. Cellulose is a very common natural polymer found in plants—and was the basis for the first plastics. The first fully synthetic plastic was invented in 1907 by Leo Baekeland. Bakelite, formed by combining the chemicals phenol and formaldehyde under heat and pressure, was a synthetic substitute for shellac, another natural product derived from insects that had many uses, including electrical insulation. Bakelite was not only a good insulator, it could be molded and shaped, it was durable, heat resistant, and importantly, it could be mass produced. Baekeland's breakthrough led major chemical companies to invest in the research and development of new polymers, and new plastics were soon making their appearance.

BIOPLASTIC

Traditional, petroleum-derived plastics do not break down by natural processes, and so plastic pollution is building up in vast amounts in the environment. Bioplastics are obtained from renewable resources and can be naturally recycled by biological processes. They are now used, among other things, in medical products, food packaging, agriculture, and horticulture. PHB, the first known bioplastic, was discovered in 1926 by Maurice Lemoigne, working with the bacterium *Bacillus megaterium*.

58 Telephone

IT IS PERHAPS TELLING THAT THE SMART DEVICES WE CARRY WITH US ARE CALLED TELEPHONES, despite the fact we use them for much more than talking at a distance. Their invention began with a man whispering into a skull.

THE HORSE DOES NOT EAT CUCUMBER SALAD

Fifteen years before Bell's triumph, the German inventor Philipp Reis built a prototype telephone and transmitted the phrase "The horse does not eat cucumber salad," which was chosen for its complexity. Reis also coined the term "Telephon" to describe how his device was for talking from afar, although it only worked in one direction.

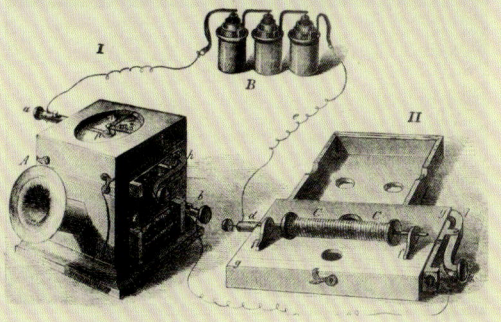

In the second half of the 19th century, many inventors raced to convert voices to electricity and back again. The Scottish-Canadian Alexander Graham Bell has the strongest claim to that innovation. Having grown up with a deaf mother and a father who developed ways of teaching deaf people to speak, Bell was interested in capturing the sound of the voice. In 1874, he built a device called a phonautograph which used the skeletal remains of a human ear—the eardrum and ossicles—to vibrate a lever which scored a pattern in soot coated on glass. Next he searched for a more practical technology that could transmit sounds along wires as patterns of electric current. The big breakthrough came on March 10, 1876, with Bell's "liquid transmitter." This device consisted of a cone with parchment stretched across the bottom like an eardrum. A needle attached to the parchment was dipped into dilute acid and wired to a battery. Bell's voice entering the top of the cone made the parchment and needle vibrate, resulting in a fluctuating current flowing through the needle and its connected circuit. The signals passed along a wire to the receiver (operated by Bell's assistant Thomas Watson in another room) and changed back into sound. While setting up, Bell spilled acid and used the device to call to his assistant, "Mr Watson, come here. I want to see you." This plea became the first words spoken on the telephone. Within months Bell was holding conversations over much longer distances, and wired telephony remained the main form of voice communication until the 2000s.

Alexander Graham Bell at the opening of the long-distance telephone line from New York City to Chicago in 1892.

59 Music Players

THE IDEA OF PRESERVING SOUND CAME TO THE FOREFRONT IN THE SECOND HALF OF THE 19TH CENTURY, when the invention of the telephone increased the understanding that sound could be captured as a pattern of vibrations.

In 1877, a young American named Thomas Alva Edison demonstrated the first machine that could record and then replay sound. This was the phonograph, consisting of a sheet of tinfoil wrapped around a grooved cylinder. A stylus cut a physical representation of sound vibrations into the foil, while another stylus could convert the undulating grooves back into waves in the air that are audible as the original sound (or something like it). The stylus and groove became the basis for recording sound for nearly a century. The tinfoil drum was abandoned, however, in favor of flat disks. Early disks were made from shellac and carried about 4 minutes of music, but from the 1930s, long-playing records (LPs) made from vinyl could contain about 30 minutes of music on each side.

In early versions of Edison's phonograph, the cylinder was turned by hand and the listener employed stethoscope tubes to hear the recording. In 1915, electric speakers were added to amplify the vibrations.

The phonographic cylinders were stored in containers. Cheaper versions used cylinders of hard wax.

Electric and electronic

Another type of player was based on technology from 1898, when the Dane Valdemar Poulsen converted sound to electrical pulses that he recorded as a magnetic pattern on a wire. When the wire was passed over another conductor, the pattern induced the same set of electrical pulses that then recreated sound inside a loudspeaker. The audio cassette, which appeared in 1963, used magnetized tape rather than wire.

A third type of technology stored sound as digital code using combinations of 0 and 1. Now stored in various formats on computers, this binary code was initially recorded onto compact disks (CDs) as a series of microscopic pits and bumps. When the disk was scanned by a laser, the flickering reflection of the laser flashed out the digital code.

MP3

As the Internet grew in the 1980s, people began to share music online but were thwarted by the amount of data required: high-quality sound needed 1.4 million bits of information per second. In 1989, the Fraunhofer Institute in Germany came up with an algorithm to remove the unheard details of sound. That compressed the overall file size by up to 12 times. The technology was eventually called Moving Picture Experts Group Audio Layer-3, or MP3.

60 Road Transportation

THE HISTORY OF MOTORIZED ROAD TRANSPORTATION IS FULL OF EVOCATIVE NAMES, FROM HARLEY-DAVIDSON MOTORCYCLES AND GREYHOUND BUSES to great automobiles like the Ford Model T or the Bugatti Veyron. With drive-ins and out of town malls, cars have changed much about society.

From the early 1800s, inventors began to develop steam-powered "horseless carriages" to carry passengers, but the vehicles were of limited use because of their bulky boilers. Progress was quicker after 1863, when the French inventor Étienne Lenoir built the Hippomobile, the first car using an internal combustion engine. It was slow and unreliable, as were its many successors. However, the German Karl Benz had considerably more success with a three-wheel Patent-Motorwagen, which was the first car that could compete with horse-drawn vehicles.

In 1901 a young American named Henry Ford formed a car company. To begin with, each Ford car was constructed by craftsmen, as was the process in other manufacturers around the world. Then, in 1908, Ford began to build the Model T using assembly-line technology so that each vehicle took only 93 minutes to build, reducing the cost of cars and putting them at the heart of modern life.

The heyday of gasoline cars came in the later decades of the 20th century, but by then people were concerned about their sustainability. In 1997, Toyota in Japan produced the hybrid Prius, which could run off either an electric battery or a gasoline engine—and the engine recharged the battery. This was followed by fully electric models, recharging at roadside points. The next evolution of "horseless carriages" will be "driverless cars."

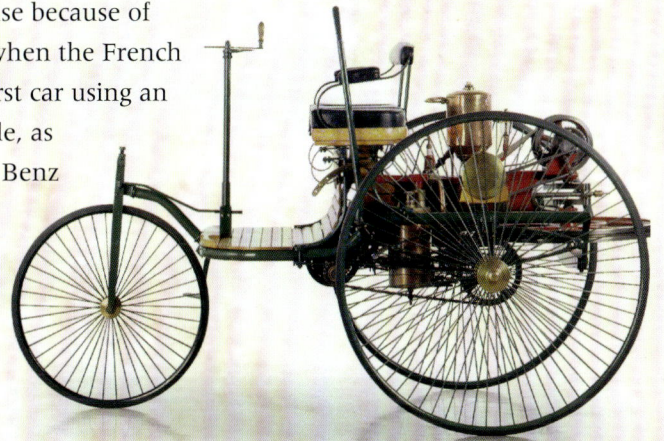

The Benz Patent-Motorwagen had no steering wheel, tires, or pedals.

STEAM CARRIAGES

The first person to put a steam engine in a vehicle was the French military engineer Nicolas-Joseph Cugnot. In the 1760s, Cugnot built two three-wheeled, steam-powered carriages (below) to pull heavy artillery weapons. Cugnot's invention was very slow, however, and not practical at all. Indeed, on its first outing, the immense and sluggish vehicle was involved in the first road traffic accident as the carriage demolished a wall.

61 Aircraft

THE SYSTEMATIC STUDIES OF FLIGHT WERE MADE IN THE 15TH CENTURY BY LEONARDO DA VINCI, who designed ingenious contraptions such as the ornithopter, a proto-helicopter, though none of his ideas were ever built.

English inventor George Cayley flew his first unmanned glider in 1804. Making improvements to his designs over the course of 50 years, he changed the shape of the wings to improve air flow, added a tail for stability, and tried a biplane design for added strength. One of Cayley's gliders made the first sustained flight with a person aboard (records differ as to whether a servant or a local child sat inside) in 1853.

Inspirations

Cayley recognized that a fixed-wing aircraft with a power system for propulsion and a tail to assist in control and steering would be the best way to stay airborne for a long time. Between 1891 and 1896, German aeronautical pioneer Otto Lilienthal built and flew a succession of working gliders, achieving a degree of lift that no other aviator could match. Orville and Wilbur Wright, cycle manufacturers from Ohio, were inspired by Lilienthal. They adopted his approach to glider experimentation and built on his aerodynamic data in their own research.

Powered flight

On October 9, 1890, French inventor Clément Ader launched his steam-powered, tailless monoplane, named the *Éole* (from the Greek god for wind), on a 50-meter (165-ft) hop. It was the first time a powered aircraft had taken off and carried a human being into the air, but *Éole* could only fly in a straight line.

Orville and Wilbur Wright spent three years experimenting with different gliders at Kitty Hawk, North Carolina, a location with the advantage of sand dunes for a soft landing. The brothers designed a wind tunnel to try out variations in wing and tail shapes. In 1902, satisfied with their design, and after hundreds of successful glider flights, they turned their attention to a propulsion system. Their first engine, designed with machinist Charles Taylor, had an output of

LIGHTER THAN AIR

On September 24, 1852, French engineer Henri Giffard piloted his airship 27 kilometers (17 miles) from Paris to the town of Trappes. Driven by a three-bladed, propeller-powered, 2.2-kilowatt steam engine, and with its direction controlled by a sail, the 44-meter (144-ft) hydrogen-filled dirigible reached a speed of 10 km/h (6 mph). It was the first passenger-carrying, powered, and steerable airship ever built, although it could only fly in calm weather.

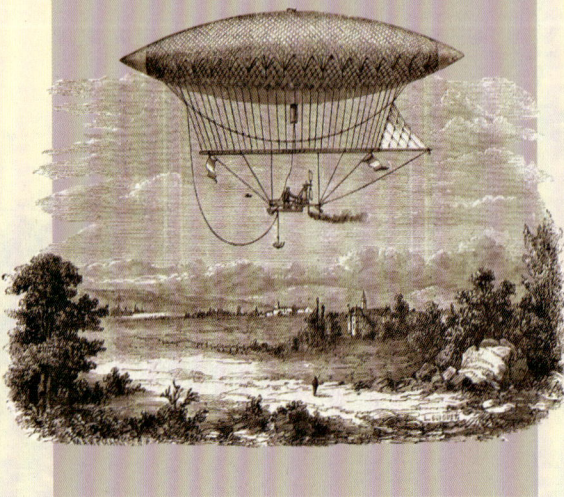

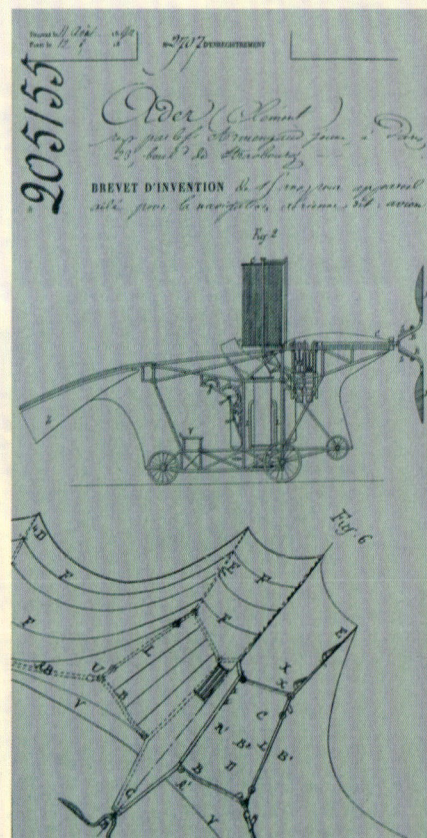

Pushed along by a rear-mounted propeller and with 14-m (45-ft) wings, the Ader Éole's *single flight reached an altitude of just 20 cm (8 inches).*

ROTARY SYSTEMS

The first helicopter flight was achieved in 1907 when Frenchman Paul Cornu succeeded in getting his twin-rotor aircraft (below) off the ground for 20 seconds, though he failed to get it down again in one piece! Igor Sikorsky built two helicopters in 1909, but they struggled to lift their own weight. The first practical helicopter was the Focke-Wulf Fw 61, which was first flown in 1936, although only two prototypes were ever built. In 1939 Sikorsky added tail rotors in addition to the main rotor, which gave stability and controlled the direction of flight.

around 9 kilowatts, about the same as a couple of modern lawnmower engines. The engine's crankcase was made out of aluminum, the first, but by no means the last, time this lightweight metal was used in aircraft construction.

Flight time

On December 17, 1903, Orville Wright took off in the brothers' *Flyer* to the north of Kill Devil Hills, near Kitty Hawk. The biplane aircraft had a wingspan of 12.3 meters (40 ft) and was 6.4 meters (21 ft) long. With pilot, it weighed 341 kilograms (752 lb). That first flight lasted just 12 seconds and covered less than 40 meters (130 feet), but later in the day Wilbur stayed in the air for 59 seconds and flew 255.6 meters (852 ft). It was the first sustained, controlled, powered flight in a heavier-than-air machine. Finding the *Flyer* hard to control, the Wright brothers worked on perfecting their aircraft design. On October 5, 1905, Wilbur took up the improved *Flyer III*, spending 39 minutes in the air.

Up, up, and away

The Wright brothers' achievements were soon providing the inspiration for a generation of aviators. By 1909, new monoplane designs were being flown by pioneers such as Louis Blériot, who made the first flight across the English Channel that year. Within a few years, pilots would be shooting at each other from the skies above Europe in World War I.

Orville Wright takes control as the Flyer *lifts off from the wooden runway to make the first airplane flight.*

In 1937, English engineer Frank Whittle contributed the jet engine to aircraft technology. (He's seen here with a scale model of the prototype.) For its size, a jet engine is capable of producing a huge amount of thrust. This innovation has allowed aircraft to reach high speeds and travel long distances.

62 The Cinema

A NUMBER OF TECHNOLOGICAL DEVELOPMENTS, SUCH AS PHOTOGRAPHY, and the discovery that still images could be combined to give an illusion of motion, led to the multi-billion-dollar movie industry.

In 1891, Thomas Edison developed the "kinetoscope," a device that gave the illusion of moving images as the viewer peered through a lens at a celluloid film strip that rapidly wound past the window. The *Cinématographe* (from the Greek for "motion"), which is now shortened to cinema, was patented by the Lumière brothers in 1895. An earlier version had been made by inventor Leon Bouly in 1892, but he failed to pay his patent fee. It was a lightweight film projector that doubled as a camera and printer. The Lumières made more than 1,000 short films of everyday life. In December 1895, in the Grand Café in Paris, the Lumières held the world's first ever commercial film screening, a sequence of about 10 short scenes, including the brothers' first film, *Workers Leaving the Lumière Factory*.

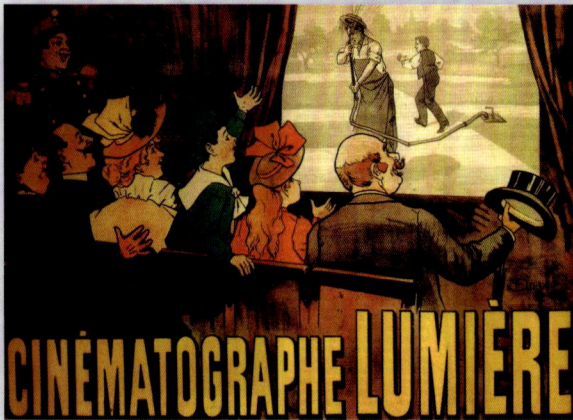

Louis Lumière believed cinema was "an invention without a future" as he thought audiences would get bored watching things they could just as easily see in real life.

63 Wireless Technology

RADIO WAVES

The waves discovered by Heinrich Hertz (below) were first named Hertzian waves, but are now known as radio waves. Hertz generated them with an electric spark, and a second spark formed in a receiver when the waves were picked up across the room. Compared to light, the waves had a longer wavelength and lower frequency, or number of wavelengths per second. Today, the unit of frequency is the hertz (Hz).

THE IDEA THAT INVISIBLE ELECTROMAGNETIC WAVES TRAVELED THROUGH SPACE was predicted by the Scottish physicist James Clerk Maxwell in 1864. Once discovered, these waves would provide a revolutionary way of communicating.

A new kind of invisible electromagnetic wave was found by the German physicist Heinrich Hertz in 1887. When Hertz was asked about possible practical uses of his discovery, he said he could see none. Hertz's views were not shared by everyone. The Serbian-American Nikola Tesla saw the possibilities of using radio waves for communication, but it was the Italian Guglielmo Marconi who in 1896 claimed a patent for wireless telegraphy. Marconi had sent a Morse code telegraph signal two miles without wires. In 1899, he transmitted signals from

Guglielmo Marconi with one of his early radio sets.

England to France, and two years later across the Atlantic Ocean to Canada. In order to transmit sounds, such as speech, the transmitter required a high-frequency electrical current to provide a continuous fluctuating radio wave. Such a device was an alternator, and in 1906 the Canadian Reginald Fessenden used his one to transmit the first voices and music by radio.

VIBRATIONS THROUGH THE GLOBE

Nikola Tesla planned a "world wireless," a global communication system that would use the planet itself as a conductor. In Tesla's vision, waves passing through the Earth and atmosphere would carry electrical energy around the planet without the need for wires. When Tesla's plans became overambitious, however, his financial backer J.P. Morgan withdrew funding. The scheme's only physical legacy was Wardenclyffe Tower, a striking radio transmitter on Long Island.

Further improvements

Radio improved quickly thanks to a series of other breakthroughs. The American Lee de Forest invented the Audion valve, which amplified weak signals so they could be powerful enough to hear, and Greenleaf Pickard invented a method of using a crystal of pure silicon to tune the receiver.

The first commercial radio show was broadcast in November 1920. Within four years, there were 600 radio stations in the United States alone. Radios now had speakers that allowed the whole family to listen to plays, comedy, news, music, and sports commentary. The American Edwin Howard Armstrong invented the FM radio system, where signals were sent by changing, or modulating, the frequency of the wave. FM receivers were able to tune to different stations with more precision. Digital radios, which went on sale in 1999, pick up signals that carry sounds as a computer code. Modern wireless communication is based on digital radio waves. This includes not only television but also cellular phones and wireless Internet.

64 X Rays

MEDICAL IMAGING, ALLOWING DOCTORS TO PROBE THE INNER WORKINGS OF THE BODY WITHOUT FIRST CUTTING IT OPEN, has transformed medicine. It all started with an accidental discovery in a darkened laboratory.

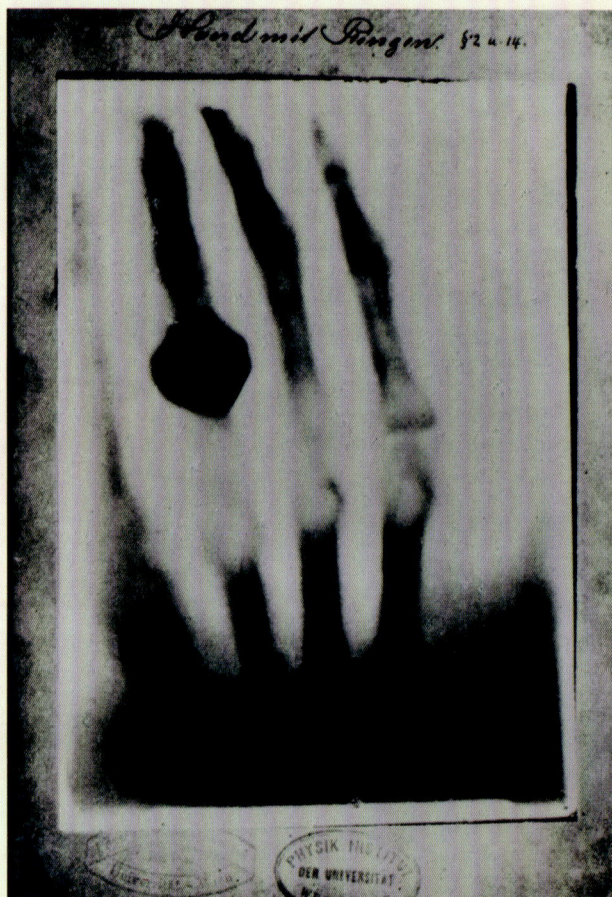

In December 1895, Wilhelm Röntgen was working with a cathode-ray tube in his laboratory, when he noticed that a nearby screen had started to glow. He concluded that a new type of ray emitted from the tube was exciting the phosphorescent crystals on the screen. Experiments established that the new ray could pass through most substances, including through the soft tissues of humans, and cast a shadow image of the bones and harder tissues on a photographic plate. In honor of his discovery Röntgen would be awarded the first ever Nobel Prize in Physics in 1901.

Röntgen labeled the mysterious rays with an X, and these X rays caught the attention of both scientists and the general public around the world, who were fascinated by the invisible ray that could lay bare the secrets of the body. Within weeks of the discovery medical radiographs were being made to guide the work of surgeons in Europe and the United States. As early as January 1896, a German medical journal reproduced a radiograph showing a glass splinter lodged in the hand of a four-year-old. In March that year, the first military radiological unit was installed in the Kaiser Wilhelm Akademie in Berlin, and soon X rays were being used by battlefield physicians to locate bullets in wounded soldiers. Double Nobel prize-winner Marie Curie invented "radiological cars," called "Little Curies." Equipped with a photographic darkroom and an X-ray machine powered by the car's engine, the Little Curies were employed on World War I battlefields. Over a million soldiers were examined using X rays in the war.

One of the first images Röntgen captured was of the bones in the hand of his wife, Anna Bertha.

MRI

Magnetic resonance imaging (MRI) scanners use a quantum phenomenon called nuclear magnetic resonance (NMR), first observed in 1937. NMR makes atoms align to a powerful magnetic field. If another energy source pushes an atom out of alignment, it will emit a pulse of radiation as it swings back into place. The MRI process magnetizes the human body with an immensely powerful superconducting magnet. This makes the hydrogen atoms in the myriad chemicals within the cells and tissues align. Next the MRI machine sends out a blast of radio waves, which make some of the atoms wobble, and thus emit a return signal. These signals are used to construct an image of the soft tissues in the target area. The first MRI scan was made in July 1977.

An early X-ray machine from 1900 is taking images of skulls. In those early days, the risks of X rays were not understood, and the patients and clinicians were not protected.

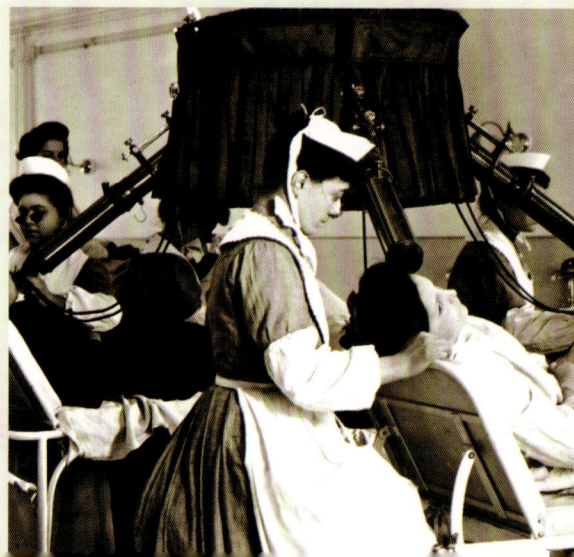

65 Vacuum Cleaners

NO ONE LIKES CLEANING, SO TECHNOLOGY THAT DOES IT FOR YOU WAS ALWAYS GOING TO BE WELCOME. The basic design of the vacuum cleaner hasn't changed much in the last hundred years or so.

An Electric Suction Sweeper from about 1908. The basic upright design has barely changed since, although disposable bags used to collect dust are often now replaced by bagless models which collect dirt in a bin.

In 1901, Royal Navy engineer Hubert Cecil Booth invented the "Puffing Billy." Powered by a six-horsepower piston engine designed by his friend F. R. Simms, the machine was pulled through the streets of London on a horse-drawn carriage. The pipes from its vacuum pump led from the roadside through the windows of the premises being cleaned. So impressive were the results, Booth was commissioned to clean the carpets of Westminster Abbey in central London, ahead of the coronation of King Edward VII in 1902.

Amateur inventor and store janitor James Murray Spangler from Ohio devised a cleaning contraption from a broom, a pillowcase, a ceiling fan motor, and a rotating brush from a carpet sweeper. The machine sucked up dirt and blew it into the pillowcase. Spangler patented this design in 1907 and left his job, opening the Electric Suction Sweeper Company. Short of cash, Spangler approached his cousin, Susan Hoover. Her husband, William Hoover, bought up Spangler's patent. The rest, as they say, is history.

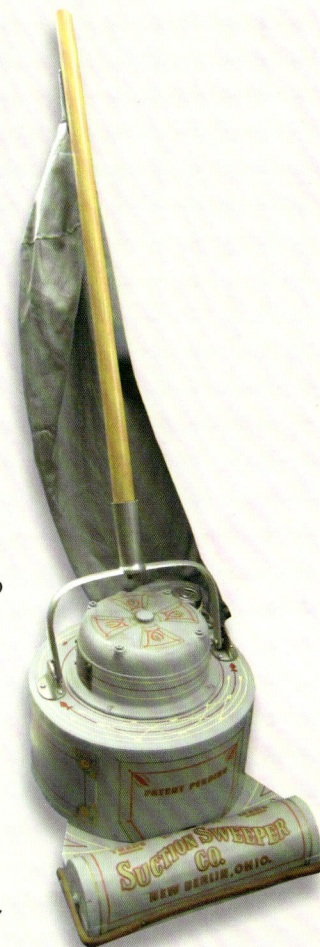

66 Fertilizers

A THIRD OF ALL PEOPLE ON EARTH AVOID STARVATION THANKS TO foods grown using chemical fertilizers. That makes them quite an invention.

Fritz Haber in his lab working on the Haber Process, which mixes nitrogen and hydrogen gas to make ammonia. The gases are heated to around 450°C (842°F), then squeezed to 200 times atmospheric pressure. The resulting ammonia is then cooled into a liquid and drained off, and used to make fertilizers, nitric acid, and explosives such as TNT.

In 1798, Thomas Malthus predicted that the rising human population would eventually outstrip the Earth's capacity to grow food. Unless a new source of nutrients could be found, the human race would starve in vast numbers. Today, however, we have the technology to eradicate famine (should we collectively choose to), and the most crucial is the ability to make chemical fertilizers, which are nutrients added to the soil so crops can grow bigger and faster. The most important fertilizers are nitrates, chemicals that contain nitrogen and oxygen. Although a common constituent of air, nitrogen is very unreactive. Nitrates are made naturally by a complex biological process, and to do so artificially was costly and tricky. However, in 1908, the German chemical engineer Fritz Haber created an industrial process, the Haber Process, for making ammonia on a large scale. This one reaction has saved billions of lives.

67 Television

FEW TECHNOLOGICAL BREAKTHROUGHS CAN HAVE HAD AS MUCH IMPACT ON EVERYDAY LIFE AS TELEVISION, but no one person can take credit for its invention; it was the culmination of a number of innovations.

John Logie Baird gives the first public demonstration of moving televised images on March 25, 1925. He showed a ventriloquist's dummy named "Stooky Bill" talking and moving (actual human faces didn't show up well on this early TV system).

People have long sought for ways of communicating over long distances, and real breakthroughs in fulfilling this desire came with inventions like Samuel Morse's telegraph in 1843 and Alexander Graham Bell's telephone in 1876. Could a way also be found to send images? In 1873, Joseph May, an electrician working on the Atlantic telegraph cable, reported that the selenium rods he was using changed their resistance when exposed to varying light levels. In 1880, Maurice LeBlanc suggested a way of transmitting moving images by using a scanning device that sent light from an image to a selenium cell that converted it into electricity. At the receiver, the signal would

MECHANICAL TV

Mechanical television uses motorized rotating disks with a spiral of holes (picture right). Light from the subject goes through the transmitter disk and hits a photoelectric cell, which converts the flickering pattern into electric pulses. The receiver disk is turning at the same speed as the transmitter (one turn for every frame—Baird's system was 12 frames a second). A lamp behind flickers according to the electrical signal, thus building up a copy of the original image.

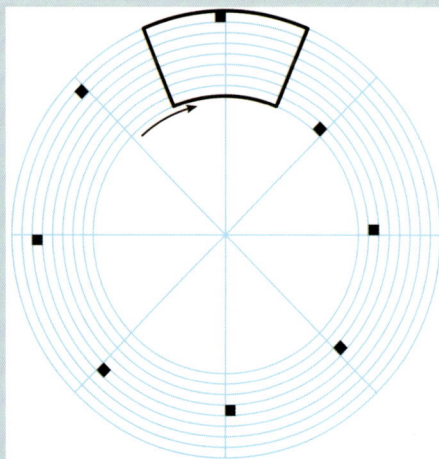

LIQUID CRYSTAL DISPLAYS

The liquid crystal display (LCD) was invented in the 1960s by George Heilmeier. LCDs have replaced cathode-ray tubes (CRT) as screens for computer monitors and televisions. A liquid crystal is a chemical that becomes opaque to light when electrified. An LCD's pixels are arranged in a grid, with each pixel having a red, green, and blue sub-pixel that can be turned on or off. Together, the pixels can form millions of color combinations.

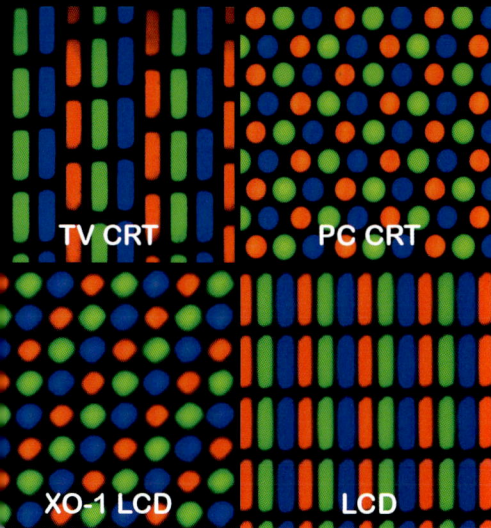

TV CRT

PC CRT

XO-1 LCD

LCD

control the light from a lamp, building up a picture of the transmitted image by the phenomenon of persistence of vision, where the brain creates the perception of motion from a series of still images. Paul Nipkow rose to the challenge and applied for a patent for his "electric telescope" in 1884. His invention became better known as Nipkow discs, and formed the basis of the first mechanical televisions systems (see box, below left).

Seeing at a distance

The term "television" first appeared in a paper written for the 1900 Paris Exhibition by Constantin Perskyi; it means "seeing at a distance." In 1909, three different television systems were built. The most promising was the one demonstrated by Georges Rignoux and A. Fournier which had a transmitting screen formed from a bank of selenium cells, each connected to a relay. As each relay was connected, it sent a signal down a wire to the receiver where a modulated light source was directed through a set of rotating mirrors, forming the image on a screen. It was the first actual television system to be built.

Nipkow's disk system was used by Scottish engineer John Logie Baird to build his successful television setup in 1925. A major problem with mechanical systems, such as Baird's, was the flickering image formed by the low frame rate. In 1923, Vladimir Zworykin devised an all-electronic television system but struggled to get further funding. By 1927, Philo Farnsworth was demonstrating another electronic system. With backing from the Radio Corporation of America (RCA) in 1931, Zworykin—inspired by Farnsworth's innovations—produced the iconoscope. This used a transmitted signal to control an electron beam, produced by a cathode-ray tube, so that it painted an image on a light-sensitive mosaic of silver globules. This was the system used in RCA's first public television broadcasts in 1939.

Vladimir Zworykin's original patent for the electronic television, which became the main TV technology until the development of digital TV and liquid crystal displays in the 2000s.

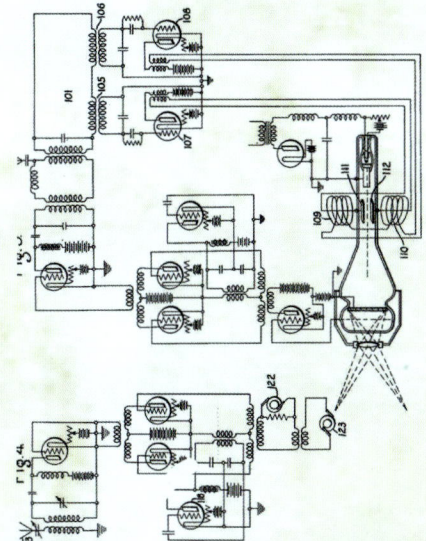

Right: Vladimir Zworykin demonstrates a 1929 model of his television set. The broadcast images were projected on a mirror on the top of the cabinet, making it possible for many people to watch at once.

68 Rocketry

THE ROCKETS OF TODAY, LAUNCHING PROBES TO THE FURTHEST REACHES OF THE SOLAR SYSTEM, had their origin in the gunpowder-powered Chinese "fire arrows" of the 10th century.

A SpaceX Falcon Heavy rocket lifts off in 2018. SpaceX is a private space launch company. In 2020, the company took over sending NASA's astronauts to the International Space Station.

Legend has it that one of the earliest rocket pioneers was a Chinese official called Wan Hu. He had the bright idea of making a rocket-powered flying chair using two large kites and 47 fire arrows. The rockets were lit, making a tremendous roar and creating huge clouds of smoke. When the smoke cleared, Wan-Hu and his flying chair had vanished. Blown to pieces or flown away? No one knows.

In theory
Russian schoolteacher Konstantin Tsiolkovsky put forward far-sighted ideas for space exploration using liquid-fueled rockets, multi-stage boosters, and space habitats. His most important work was *Exploration of Outer Space by Means of Rocket Devices*, published in 1903. Although he never actually built a rocket, his vision of the future of space exploration, plus the mathematics of flying into orbit known as the rocket equation, earned him the title of father of modern astronautics.

Flying machines
American Robert Goddard carried out experiments in rocketry aimed at reaching higher altitudes than were possible using balloons. He became convinced that liquid fuels would perform better than solid propellants, although there were technical problems to be faced, such as building fuel tanks and combustion chambers. Overcoming the difficulties, Goddard flew the world's first liquid-fueled rocket at Auburn, Massachusetts, on March 16, 1926. The rocket flew less than 60 meters, but local

MULTISTAGE ROCKETS
Multistage rockets were developed in theory by Tsiolkovsky (above), Oberth, and Goddard. This is a type of rocket that is effectively made up of one or more separate rockets either stacked on top of one another or attached alongside each other. The development of more and more powerful rockets in the years after World War II was spurred more by the need to produce a delivery system for the atomic bomb than scientific curiosity. Russian rocket engineer Sergei Korolev designed the multistage R7, with several small engines clustered around a central core, with the aim of delivering a 2-tonne hydrogen bomb. It was an R7 that put Sputnik, the first artificial satellite, into orbit on October 4, 1957.

fire marshals were so alarmed by Goddard's activities they barred him from further experiments. Goddard moved to New Mexico where, among other things, he developed the first gyro-controlled rocket guidance system and a parachute recovery to get his equipment safely back to Earth.

Flying bombs

A third great space pioneer was the German Hermann Oberth. In 1923 he published *The Rocket into Interplanetary Space*, which became the inspiration for rocket societies that were formed across the world. Banned from having long-range artillery after World War I, the German government was interested in the possibility of using rockets as weapons, and in 1929 it sponsored Oberth's work. He was aided by an 18-year-old student, Wernher von Braun. In 1937, an engineering group led by von Braun began to produce a series of liquid-fueled rockets. This led to the V-2, the world's first long-range ballistic missile. Just 14 meters (46 ft) long, the V-2 was small in comparison to many of today's rockets. Its engines burned a mixture of liquid oxygen and alcohol to produce a colossal thrust capable of delivering a massive bomb to targets 320 km (200 miles) away at three times the speed of sound. In the late stages of World War II in Europe, V-2s caused huge damage. German rocket engineers and V-2s captured at the end of the war formed the basis of the USA's and the USSR's post-war missile development and space exploration programs.

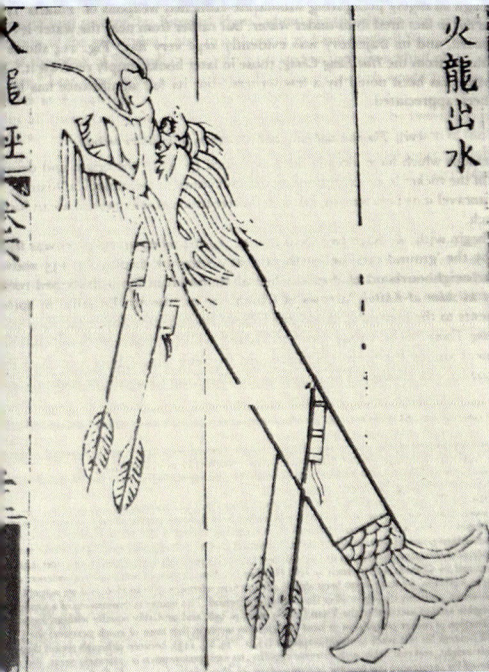

Huo Long Chu Shui (Fire Dragon Out of the Water) is an ancient Chinese weapon where a bamboo tube packed with explosives was launched at the enemy by four rockets, which ignited the flying bomb.

REUSABLE BOOSTERS

Getting into space is hugely expensive, not least because rockets are used only once. The space shuttle was the first attempt at a reusable spacecraft, but the cost of refurbishment after every flight meant it cost as much as a single-use spacecraft. In April 2016, SpaceX launched the Falcon 9 reusable rocket booster on a mission to the International Space Station. The rocket successfully flew back down to the launchpad, becoming the first rocket to touchdown vertically on a ship at sea. It was refueled and flew again in March 2017.

69 Electronics

ELECTRONIC TECHNOLOGY USES THE FLOW OF ELECTRONS THROUGH CIRCUITRY TO CONTROL DEVICES, PERFORM CALCULATIONS, and process inputs into outputs. This involves using high-speed switches called transistors.

Before the invention of the transistor, the first electronic devices used vacuum tubes (themselves a product of light bulb tech), which had been modified in 1907 by American Lee de Forest to amplify the weak electrical signals formed by early radio receivers. Vacuum tubes did this by having three electrodes, and that meant the device could be switched on—so a current flows through from one electrode to a second—and off again by applying a current to the third electrode. Here were the switches needed for electronics, but the tubes had serious drawbacks. The glass was vulnerable to cracking and they also generated a great deal of heat, so burned out quickly.

Semiconductors

Engineers looking for an alternative to the vacuum tube turned toward the growing field of semiconductor research. A semiconductor is a material that has a conductivity somewhere between that of a conductor, such as a metal, and an insulator, such as plastic. In 1926, physicist Julius Lilienfeld applied for a patent covering what could be considered a transistor since in his device the flow of an electric current was controlled using a semiconductor. However, it proved impossible to build. At the end of World War II, John Bardeen, William Shockley, and Walter Brattain at the Bell Telephone Laboratories (Bell Labs), New Jersey, were researching the behavior of germanium crystals as semiconductors that could replace vacuum tubes as switches or relays for telephones networks. In 1947, Bardeen and

LOGIC GATES

Computers store and manipulate numbers in binary code (zeros and ones) by switching transistors on (one) and off (zero). Calculations are carried out using logic gates, which are basically circuits with one or two inputs and an output. The simplest are AND, OR, and NOT. An AND gate outputs 1 if both its inputs are 1; an OR gate outputs 1 if either of its inputs is 1; NOT gates, which have a single input, always produce an output that's the opposite of the input.

The first transistor would cover most of the palm of your hand. The sizes of today's descendants of this crude prototype are measured in billionths of a meter.

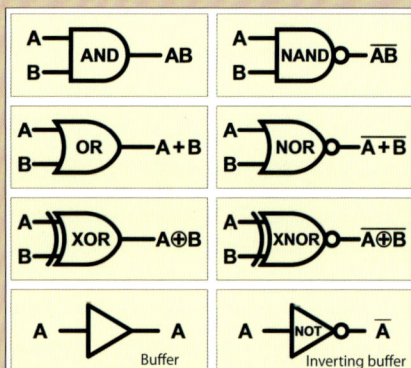

SEMICONDUCTING CRYSTALS

Semiconductors began with silicon and germanium, which are the most abundant metalloids, or semi-metals. When pure, they are poor conductors, but by adding other elements—doping the crystals—engineers create semiconductors that can switch from insulator to conductor, ideal for electronic switches. Interest in silicon and germanium crystals, and those of other similar elements, grew with radio technology, since pure crystals make good receivers. This work reached a peak with radar systems created during World War II, and the pure crystals made then formed the foundations of the first transistors.

Brattain invented the first "point contact" transistor amplifier. They discovered that when two gold foil contacts were applied to a crystal of germanium, an electric current applied to one contact was boosted by the germanium so the strength of the current flowing through the other contact was greater. Shockley found ways to improve on their design and all three were jointly awarded the 1956 Nobel Prize in Physics for the invention of the transistor.

Computing power

The transistor became a breakthrough technology, an invention that continues to change the world daily. It revolutionized communications but also computing—so much so that these two revolutions are now merged together and continue to play out today.

No longer dependent on unreliable, bulky, heat-generating, and power-hungry valves, computers could now be built that not only took up less space and used less power, but also were immeasurably more efficient. The most common use of the transistor is as a switch. It can turn a current on or off thousands of times a second. In 1958, a series of transistors set up in various configurations were joined in an integrated circuit, the brainchild of Jack Kilby at Texas Instruments. Up until then, there was a limit to how small a transistor could be made as it had to be connected up to the other components in the circuit. Kilby's insight was that all parts of the circuit could be made from a single crystal of silicon, or microchip. The first "solid circuit" produced was the size of a pencil point. Robert Noyce at Fairchild Semiconductor managed to get the first patent for a microchip in 1961 but, today, both men are acknowledged as having independently developed the idea.

The Intel 4004, released in 1971, was the first microprocessor, where the functions of a central processing unit were located on a single chip. The transistors and the connections between them were etched directly onto a chip of pure silicon cut from a larger wafer.

The insides of the Altair 8800 of 1974, the world's first personal computer. Microchips for the processor and memory filled several motherboards.

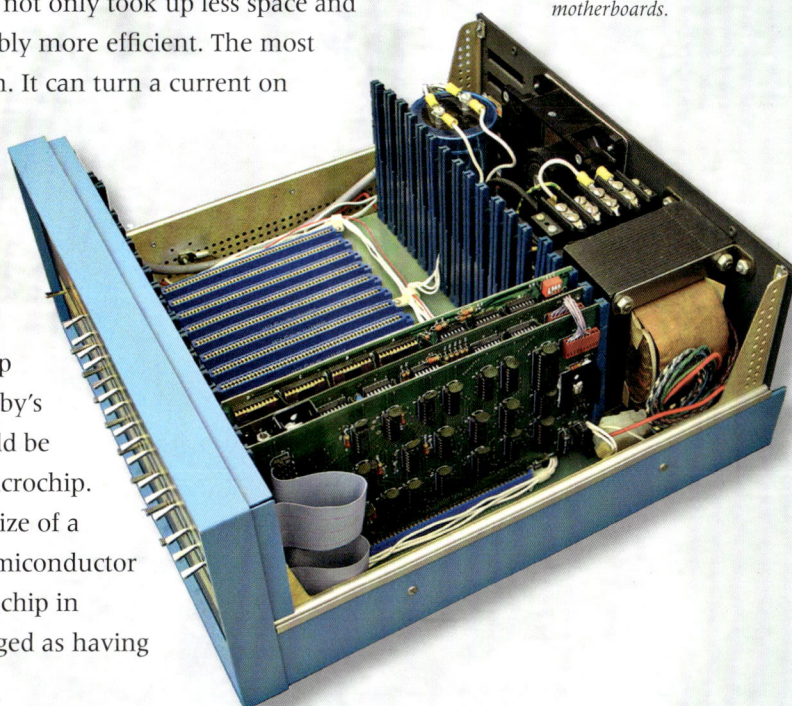

70 Holograms

SINCE THE DEVELOPMENT OF TELEVISION AND CINEMA, IT SEEMED OBVIOUS THAT THE NEXT STEP IN VIDEO TECHNOLOGY would be to view scenes in three-dimensional holograms. However, this step has proven hard to take.

The patent for holography was taken out in 1947 by Dennis Gabor, who worked with beams of electrons to demonstrate the principle. Holograms that could be seen with the naked eye needed lasers, which arrived in the 1960s. Our eyes perceive shapes from the light reflected and scattered from the shapes' surfaces. A hologram works by splitting a laser beam in two. One reflects off the object onto a photographic detector, while the second beam is directed unaltered to the detector. Both beams create an interference pattern. Light shining through this capture pattern will recreate the light waves coming from the original object, which the eyes perceive in three dimensions. The challenge for holography is how to capture moving objects effectively and project them into a space where the benefits of their three dimensions are clear. Without that advantage, 3D TV (or even basic 2D) is easily more effective.

A holographic setup uses two-way mirrors and lenses to direct the laser's beams.

71 Solar Power

JUST A FEW YEARS AFTER CHANGING THE WORLD WITH THE TRANSISTOR, engineers at Bell Labs did the same again by inventing a new power generation system.

The link between light and electricity had been made in the 19th century (see box, right) but crude attempts at using sunlight as a power source met with failure. The development of n- and p-type semiconductors during the invention of transistors at Bell Labs in 1947 led to further innovations in solar power. In 1954 three inventors—Gerald Pearson, Calvin Fuller, and Daryl Chapin—developed the first efficient silicon-based solar cells for the conversion of solar energy into electricity. Their invention was given the seal of approval by NASA, and Explorer 6, launched in 1959, had four silicon solar arrays that folded out once in orbit to provide power for months in space.

The Bell "Solar Battery" found limited use back on Earth, where oil was a very cheap source of energy. In 1974, only six homes in North America were heated by solar power. Experts

Left: The first solar panel, or Solar Battery as it was called then, was installed in the U.S. state of Georgia in 1955 to power the telephone network.

now forecast that by 2050 solar power will be the world's biggest single source of electricity—at 27 percent of the total—and cheaper than electricity generated by thermal power plants.

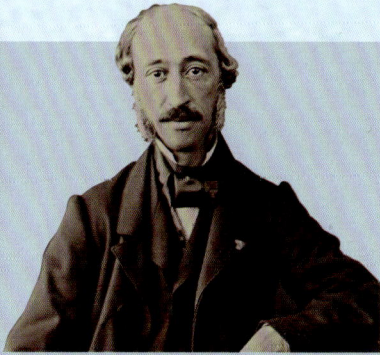

THE PHOTOVOLTAIC EFFECT

The original solar panel, and many solar power generation systems, rely on the photovoltaic effect. This is when light hitting a substance results in a current flowing through the material. The effect was discovered in 1839 by French physicist Edmond Becquerel (whose son Henri discovered radioactivity 57 years later). The effect works by photons transferring energy to molecules. These release free electrons which can form a current. This process is especially effective using layers of semiconductors where one layer has a lack of electrons (p-type) and another has an excess (n-type).

Direct or focused

There are two main kinds of solar power: photovoltaic (PV) and concentrated solar power (CSP). A photovoltaic solar farm works like the original Solar Battery. It converts the Sun's rays into electricity. Each panel contains two layers of a photovoltaic material, usually silicon. The first layer is "overloaded" with electrons (n-type). When photons of light from the Sun hit it, the electrons gain energy, become excited, and move to the second layer, which has the capacity to take on more electrons (p-type). This flow of electrons becomes an electric current. PV solar power can now be generated on any scale. Pocket calculators are powered by a single tiny cell, while Solar Impulse 2 is an aircraft entirely powered by photovoltaic cells that cover the top of its fuselage and wings. (In 2015, the aircraft flew nonstop from Japan to Hawaii, the longest-ever journey by a solar-powered plane.)

CSP produces electricity using parabolic mirrors that concentrate sunlight onto a circuit of volatile liquid, which evaporates into a high-pressure vapor that then drives a turbine generator. Alternative designs focus the light onto a single solar furnace that creates steam to drive turbines.

Beams of reflected sunlight are focused on a heat tower, where the concentrated light gets hot enough to be a heat source for fuel-less furnaces that reach 3,500 °C (6,330 °F) and melt most metals.

Vast areas of the world's deserts, where the skies are normally clear and sunny, are being given over to solar farms using thousands of photovoltaic arrays.

72 Containers

IN 1937, NORTH CAROLINA TRUCKER MALCOLM MCLEAN WATCHED HIS TRUCK BEING UNLOADED ONTO A SHIP AND THOUGHT, "Wouldn't it be great if the whole trailer could be picked up and loaded aboard?"

The standard 12.2-meter (40-ft) shipping container can be handled by cranes in every port. There are more than 20 million of them in constant motion around the world.

In 1956, McLean's idea of an "intermodal container" become a reality. For the system to work, the containers are standardized and have an ingenious device called a twistlock. This slots into the corner fitting of a container, locking it into a stack, while being easy to unlock and remove. A crane driver can simply reach down, latch onto, and unlock a container, lift it, and release it, speeding up the process of loading and unloading. McLean's containers have since been modified for use on trains, and there are refrigerated variants for transporting perishable goods.

73 Data Storage

AS WELL AS A PROCESSOR, DATA INPUTS, AND OUTPUTS, A COMPUTER NEEDS A WAY OF STORING CODE. There have been many ingenious versions of computer memory using sound, light, and magnetism to hold a record of 1s and 0s.

TORSION WIRE MEMORY

In the 1960s, one of the more reliable forms of memory was torsion wire, where a long wire was fed into a coil (below) and twisted as it passed through electromagnets that flickered on and off according to the binary code. A 1 made a twist, and a 0 did not. A single wire coil that was about 30 cm (1 foot) across could hold an entire kilobyte of data.

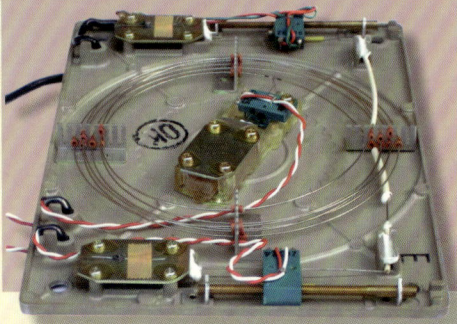

In the early digital computers of the 1940s, data was "memorized" in an ingenious system called a mercury delay line, which had been invented by J. Presper Eckert. Electrical pulses of data representing the binary code were converted into sound waves and sent through long tubes of mercury. The speed of sound in mercury is very fast, so these sound signals surged through the liquid until they hit a quartz crystal at the far end. This crystal had the property of releasing an electric current when squeezed physically. The impact of the dense mercury converted the sound wave back into an electrical signal, which was then relayed as sound into the mercury. And so data could be bounced backward and forward inside the mercury for as long as was needed. Delay lines like this were effective and returned stored data to the processor in a fraction

The IBM 305 RAMAC computer released in 1956 was the first commercially available computer system to include hard disk storage—although it took more space than today's versions, which are about the size of a side plate.

of a second, but they were cumbersome. A delay line 1.5 m (5 ft) long could only hold 576 bits, or the code for 32 letters.

To store more data in smaller spaces, computer scientists turned to electromagnetic systems, where digital code is stored as patterns of magnetized and demagnetized sectors. Initially this was done using tape covered in iron oxides. Later this developed into disks of magnetic material. First introduced in 1956, this mode of data storage eventually became known as a hard disk. This differentiated it from the "floppy" disk, which was a thin and flexible portable version of magnetic storage. By contrast, the hard disk was a permanent component within the computer and is still largely the main source of storage. Each disk has several platters, all spinning at up to 15,000 revolutions per minute. A spindle, which is a bit like a high-tech version of the needle of a vinyl turntable, can swing back and forth to position an electromagnetic read-write head in any location on the platter. To save data, the head magnetizes sectors of the platter—and de-magnetizes others if necessary—to make a pattern that represents 1s and 0s. To read data from the disk, the spindle swings across a platter, and the magnetized sectors induce a current in the head, while the demagnetized areas do not, and this pattern of currents forms an input of 1s and 0s.

Handheld devices such as smartphones and tablets are too small and slim for hard disks. Instead they use flash memory, invented in 2000 and named because it is not only smaller but faster than a hard disk. It stores data using components similar to those on a microchip.

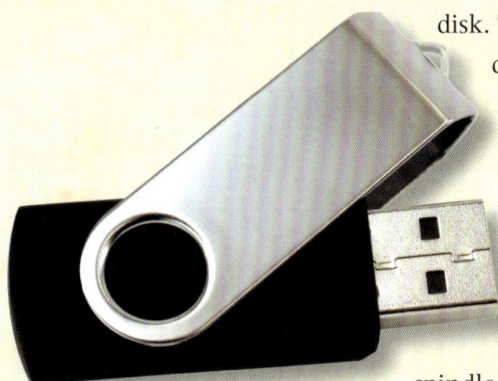

A thumb drive is a compact storage device that contains flash memory.

OPTICAL STORAGE

In 1995, the Digital Versatile Disk, or DVD, became the latest design of optical disk, so-named because the data on the disk is read using laser light. The optical disk had been invented in the 1960s. The disk's surface has a spiral containing hollows, or pits. Spinning the disk makes a laser run along this spiral, reflecting off the flat surface but not off the pits, creating a corresponding flickering pattern. A detector equates the on and off laser reflections as the 1s and 0s of computer code. CDs and earlier optical disks had 800-nanometer pits. A DVD crams in more data by having 400-nm pits in two layers. Blu-Ray disks, which have superseded DVDs, have pits only 150 nm across.

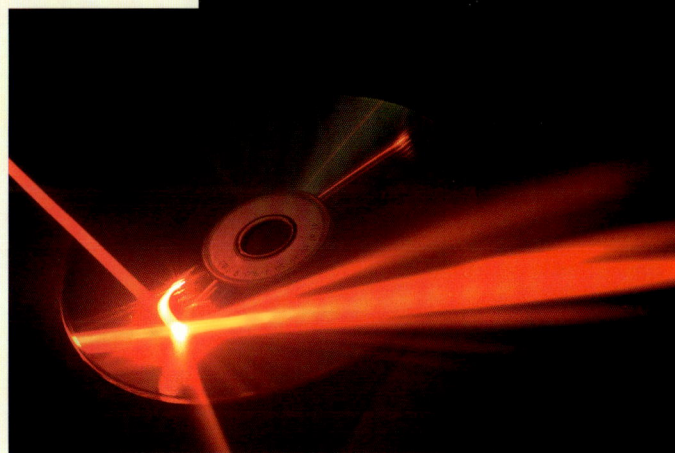

74 Lasers

A LASER IS A BEAM OF LIGHT CREATED BY EXCITING THE ATOMS INSIDE CRYSTALS OR VAPORS. Its power and precision means that lasers have found uses in industry, medicine, and entertainment.

A helium-neon laser is formed when the electrified helium gas collides with the neon, producing high-energy red light.

In 1960, Theodore H. Maiman built the first laser using a ruby crystal with its ends coated in mirrors of silver. Light was shone on the ruby, which excited the electrons in the crystal. The excited electrons gave out more photons, which bounced backward and forward between the mirrors, exciting more atoms and steadily accumulating.

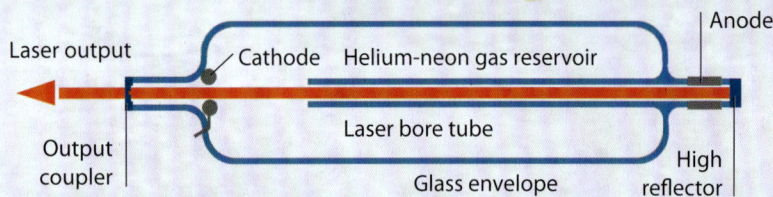

When the photons were released from the crystal, they formed a powerful beam of red light. The basic principle of lasers (or Light Amplification by Stimulated Emission of Radiation devices) has remained the same since.

Laser output
Cathode   Helium-neon gas reservoir   Anode
Laser bore tube
Output coupler
Glass envelope
High reflector

75 The Internet

IN A LITTLE OVER HALF A CENTURY THE INTERNET HAS TRANSFORMED FROM A MILITARY COMMUNICATIONS NETWORK into an all-pervading dataspace that sometimes seems to have found its way into every part of our lives.

If the Internet started anywhere it was at the research division of the U.S. Department of Defense, the Advanced Research Projects Agency (ARPA, which later became the Defense Advanced Research Projects Agency, DARPA). In 1962, J.C.R. Licklider, an ARPA scientist, suggested connecting computers through a distributed network with multiple connections. That setup would ensure active communications were maintained in the event of a nuclear attack, which would more than likely cut dedicated point-to-point connections. In 1969, ARPA created ARPANET, a computer network that linked mainframes at universities, government agencies, and defense contractors around the country.

Making the connection

The first four nodes of ARPANET were UCLA's Network Measurement Center, Stanford Research Institute (SRI), University of California, Santa Barbara, and the University

EMAIL

The first major network application, electronic mail, or email, was initiated in 1971 to provide ARPANET developers with an easy way of coordinating their activities. Ray Tomlinson wrote the basic email send and read software, including the @ symbol. In July 1972, MIT researcher Lawrence G. Roberts wrote the first email program that was able to list, read, file, forward, and respond to messages. Email really took off, and the potential for "people-to-people" communication was a major reason that the network grew so much.

of Utah. On October 29, 1969, the first message was sent over the proto-network: "lo." Student Charles Kline had attempted to type "login," but the system crashed on the "g!" Over the next few years, more and more computers were linked into ARPANET. In December 1970, the initial ARPANET protocol was completed. Called the Network Control Protocol (NCP), it allowed network users to begin developing applications to run on the network. The first public unveiling of the new network technology was in October 1972 at the International Computer Communication Conference (ICCC). Demonstrations included interactive chess games, geography quizzes, and an air traffic control simulation. A key component of the Internet is the concept of open architecture networking. Individual networks can be designed and developed according to their users' needs and then linked into other networks by the "internetworking architecture." It was a concept introduced by Bob Kahn in 1972.

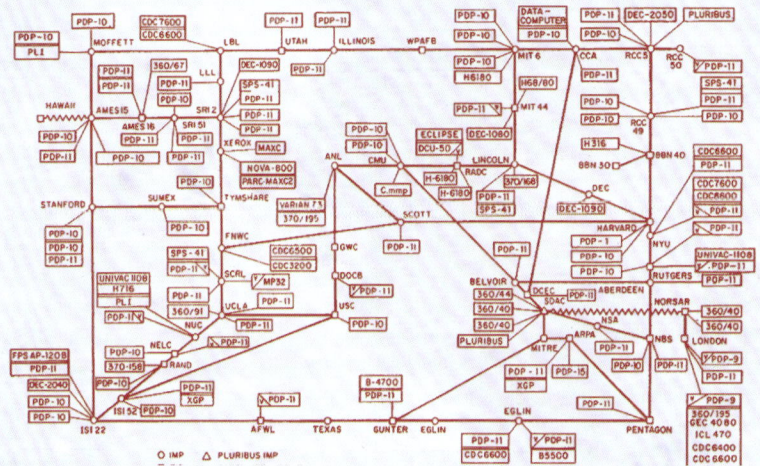

Vinton Cerf (left) and Robert Kahn (center) receive the Presidential Medal of Freedom from U.S. President George W. Bush. Kahn and Cerf were the lead developers of the TCP/IP system used by the Internet.

A map of ARPANET in 1977 covering most of the mainland of the United States and including satellite links to Hawaii and London.

The ARPANET was developed at the height of the Cold War to maintain communications with parts of the critical early-warning radar network, such as the RAF Fylingdales listening station in England.

Network of networks

The starting date for the Internet is often given as January 1, 1983. This was when ARPANET adopted a new protocol called Transfer Control Protocol/Internet Protocol (TCP/IP), which allowed all kinds of computers on different networks to connect. The Internet was used to transfer files directly from one computer to another, distribute emails, and host bulletin boards. The World Wide Web would not appear until the 1990s.

PACKET SWITCHING

Lawrence Roberts, chief scientist at ARPA, linked two computers using a telephone line and transferred digital data using a process called packet switching, a technology that still underlies Internet traffic today. A large message is split into thousands of individual, equal-sized "packets" of around 1,500 bytes apiece. Each packet takes a different route through the network, and they are only reassembled in the right order at the final destination. It is very reliable and efficient, allowing data to be sent securely, even if the network is damaged. The theory of packet switching was developed by Roberts and Leonard Kleinrock. In Britain, Donald Davies at the National Physical Laboratory independently developed the same protocol.

76 Video Games

THE VIDEO GAMES MARKET IS HUGE, WORTH ALMOST $200 BILLION. It is estimated that, globally, there are over 2.5 billion video gamers—around a third of the world's population.

The earliest example of an electromechanical game machine debuted in September 1940 at the World's Fair in New York. Based on an ancient, math game called Nim, the Nimatron was the invention of nuclear physicist Edward Uhler Condon. Reputedly 50,000 people took on the challenge at the Fair, with the computer winning 90 percent of the time.

The first video game is likely to have been the creation of physicist William Higinbotham in October 1958. Wanting to develop more interesting exhibits for visitors' day at Brookhaven National Laboratory in New York, he adapted a small computer and an oscilloscope display to produce a game called Tennis for Two. The "ball" was a dot bouncing back and forth across the screen, but it quickly became the most popular exhibit, with people standing in line to get a chance to play.

The first commercial game system designed for home use was the "Brown Box," invented by engineer Ralph Baer and produced in 1967. Powered by vacuum tubes, the Brown Box was connected to a television and could be programmed to play games like ping pong. It even featured accessories such as a "lightgun" for target-shooting games. The Brown Box was licensed to Magnavox, which, in 1972, repackaged it as the Magnavox Odyssey, the first true home video game system. The transistorized Odyssey had no microprocessor, and its very limited graphics, basically a few small, white blocks and a vertical line, were augmented by translucent color overlays for the TV screen.

Also in 1972, Allan Alcorn of Atari developed a coin-operated arcade game called Pong. This was so popular that, in 1975, it became available as a home console game, called, imaginatively, Home Pong. Pong was essentially an improvement on the Brown Box's ping pong game, and its success spurred the development of the home games market. The Fairchild Channel F, released in 1976, and the Atari 2600 Video Computer System, released in 1977, were the first games consoles to feature game software stored on programmable, read-only cartridges that were inserted into slots on the console.

THE PLAYSTATION

The PlayStation games console was first released by Sony in 1994. It was the first console to move away from cartridges and instead use optical disks for data storage. Games were designed to operate with a controller, in contrast to PC games that used a keyboard or mouse. Later PlayStation models have integrated with Internet technology, so games are downloaded and streamed, and players can connect with many others to join multi-player games.

A game of Pong was a very exciting prospect for families in the 1970s, but today's gamers would be somewhat underwhelmed.

Modern games consoles—and computers—rely on a dedicated processor, the GPU or graphical processing unit, to render complex and fast-changing graphics. That's hot work, and so GPUs need fans.

77 The GUI

IT MIGHT BE DIFFICULT TO IMAGINE IN A WORLD OF ICONS, EMOJIS, AND STATE-OF-THE-ART COMPUTER GRAPHICS, but early computers were nothing special to look at—and consequently, they were hard to control and use without a considerable level of expertise.

That changed in the late 1960s thanks to a visionary academic named Douglas Engelbart, who at the time worked in the Augmentation Research Center (ARC) at the Stanford Research Institute (later SRI International) at Menlo Park in California. In the last weeks of World War II, Engelbart had read an article about information technology by another pioneer, Vannevar Bush, who had proposed a huge postwar initiative to gather and record the world's knowledge. Inspired by Bush's approach, Engelbart was one of the first people to appreciate the capacity of computers to improve life for everyone on the planet by providing new ways to consider serious problems.

Engelbart committed his career to making computers easier to use. He and his associates at ARC led the development of the first graphical user interface, or GUI (pronounced gooey). In 1968, he displayed it for the first time at an event that changed the course of computing, now known as the "Mother of All Demos." The team from ARC demonstrated a display screen that featured the now-familiar on-screen windows, vector graphics (at that time in grays and whites), and clickable icons and links. Not only that, the user controlled the cursor on the screen with a device of Engelbart's invention, the computer mouse.

Building on Engelbart's breakthroughs, the electrical engineer Alan Kay further refined the GUI when he worked at the Palo Alto Research Center run by Xerox. It was Kay who developed the essentials of the modern GUI, making the screen a virtual desktop where the contents and applications of the computer are displayed. Windows showing files or documents overlay each other and shuffle back and forth. Commands are selected from drop-down menus at the top of the screen, and files are organized in folders just as they might be in a real-life office cabinet. Xerox did not see GUI technology as a commercial prospect, but when Steve Jobs and Steve Wozniak of Apple visited PARC, they certainly did!

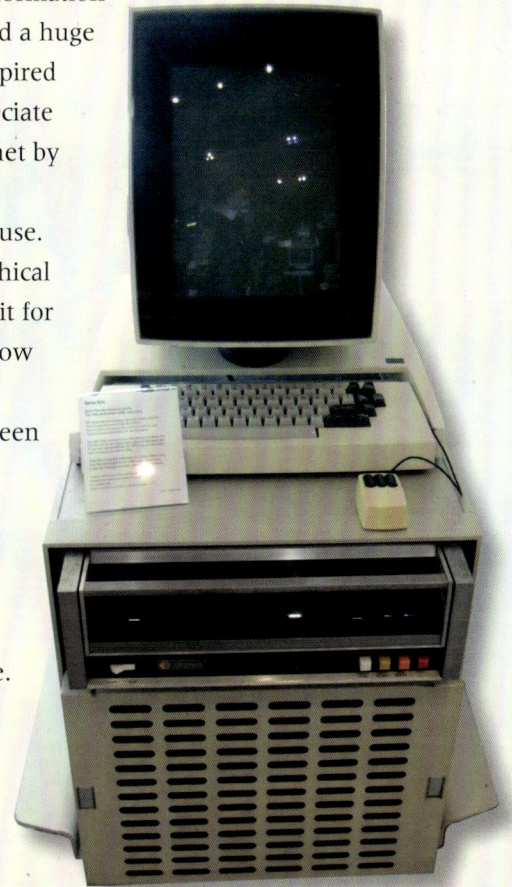

THE MOUSE

To activate the first GUI, Engelbart designed a small wooden box with wheels, joined to the computer by an electric cord. The movement of the box was tracked by a cursor on screen. The wire originally came out of the front of the box, but when that proved cumbersome, it was moved to the back, where its appearance prompted Engelbart to name it a mouse's tail. The early versions were made by Bill English of SRI from Engelbart's ideas, and later the wheels were replaced by a ball (as seen below). The mouse would eventually be popularized—and commercially exploited—when Apple included it with their early home computers.

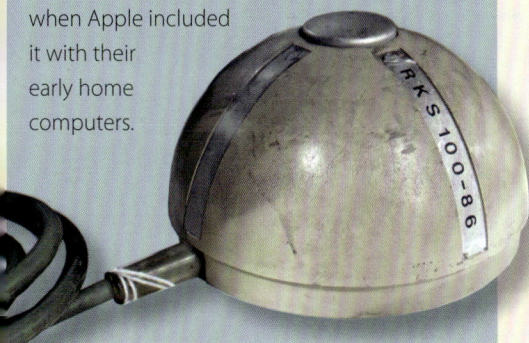

The Xerox Alto, a prototype personal computer from 1973, had a GUI and many of the features of today's personal computers.

78 Touchscreens

WHEN THE FIRST TOUCHSCREENS APPEARED IN THE 1970S AND 1980S, THEY SEEMED ALMOST MAGICAL to people used to typewriters and even computer keyboards. Those first screens laid the basis for all tablet computers.

One of the challenges faced by the designers of the earliest computers was to find the easiest way to operate the devices. One answer was the keyboard, another the computer mouse. A third was simply touching the screen to highlight data or instructions. This method was developed when an instructor at the University of Kentucky, Sam Hurst, began thinking about alternative ways to communicate with a computer, and came up with a tablet-like screen activated by being pressed with a stylus. Elographics, the company that Hurst founded, came up with a transparent screen that would sit over a computer.

Pressure-based

In 1977, Elographics came up with the first modern touchscreen. This used layers of transparent conductive material carrying an electrical charge. When the layers were pressed together, they closed a switch in the circuit and registered a location on the screen. This is called a resistive touchscreen, and it is a tough design still used today in public touchscreens. More modern personal devices use capacitive touchscreens, where the electrostatic charge of the screen is changed by the natural electrical conductivity of a finger. This type of touchscreen will not work if a person is wearing gloves or has wet fingers.

Smart devices

Touchscreens grew increasingly common with the emergence of the first smartphones and tablets in the 1990s. As with other computing breakthroughs, it took the marketing muscle of Apple to make them everyday items, first with the iPhone in 2007 and a few years later with the iPad. Today, modern touchscreens are everywhere. We use them to check in at doctor's surgeries and airports, to withdraw money from ATMs, or even to order food from menus in fast-food restaurants.

Capacitive touchscreens use a transparent conducting layer that sits under the protective outer glass or plastic.

TABLET COMPUTERS

Touchscreen technology lay behind the development of the tablet computer, initially called a personal digital assistant, in the 1990s. A tablet (below) is operated by a capacitive touchscreen and has a graphical user interface based on icons and applications, or apps, specially designed to be operated by a single finger movement. The popularity of tablets grew exponentially after Apple released the first iPad in 2010, since when they have become available in a range of sizes, and also in a version that can easily be turned into a laptop with the addition of a keyboard.

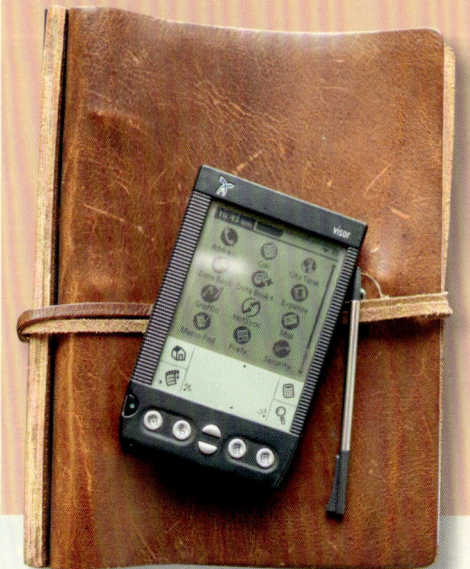

79 Encryption

IN 1977 THREE MIT COMPUTER SCIENTISTS CREATED A PIECE OF MATH THAT WAS RELATIVELY SIMPLE if you knew the two numbers concerned, but fiendishly difficult to figure out if you did not. This one-way calculation forms the basis of Internet security.

The Internet carries information as digital code, but in the simplest terms, a code that everyone understands. Private information needs to have that code converted into an encrypted form which is meaningless to those who do not have the key to decipher it. The difficulty is that in order to encrypt and decipher private messages, two communicators, e.g. Alice and Bob (or at least their computers), must share the key. They never meet, and so share the key through the Internet in full view of everyone. The 1977 breakthrough, called the RSA algorithm after its inventors Ron Rivest, Adi Shamir, and Leonard Adleman, allowed Alice and Bob to exchange a public key but still keep their information very secure. Bob wants to send Alice some money, so asks her to send her public key. This is a very big number produced by multiplying two big prime numbers. The primes are never made public; only Alice knows them. Bob does not need to know them. He uses the public key to encrypt his financial details and sends that coded information to Alice—again this is in full view of any snoopers. Alice uses her private key, which is based on the initial primes used to create the public key, to decrypt Bob's message. Now for the clever bit. The primes are the only factors of the public key, and so any snooper need only figure out what those primes are to crack the code. But, there is no way of predicting primes, so the snooper would have to try every number. It would take two years for the world's fastest computer to find the primes used in the most basic encryption, and many centuries for the most secure.

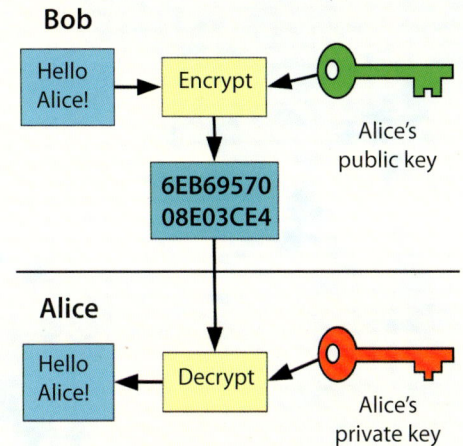

Bob

Hello Alice! → Encrypt ← Alice's public key

6EB69570 08E03CE4

Alice

Hello Alice! ← Decrypt ← Alice's private key

The process is simple if you have the right numbers, and nigh on impossible when you do not!

Online encryption uses a lot of random numbers to help obscure the identities of users active on the network. No computer can be programmed to produce truly random numbers; the program would always create some kind of pattern. Instead, random number generators use the randomness in natural phenomena, such as the shapes in lava lamps.

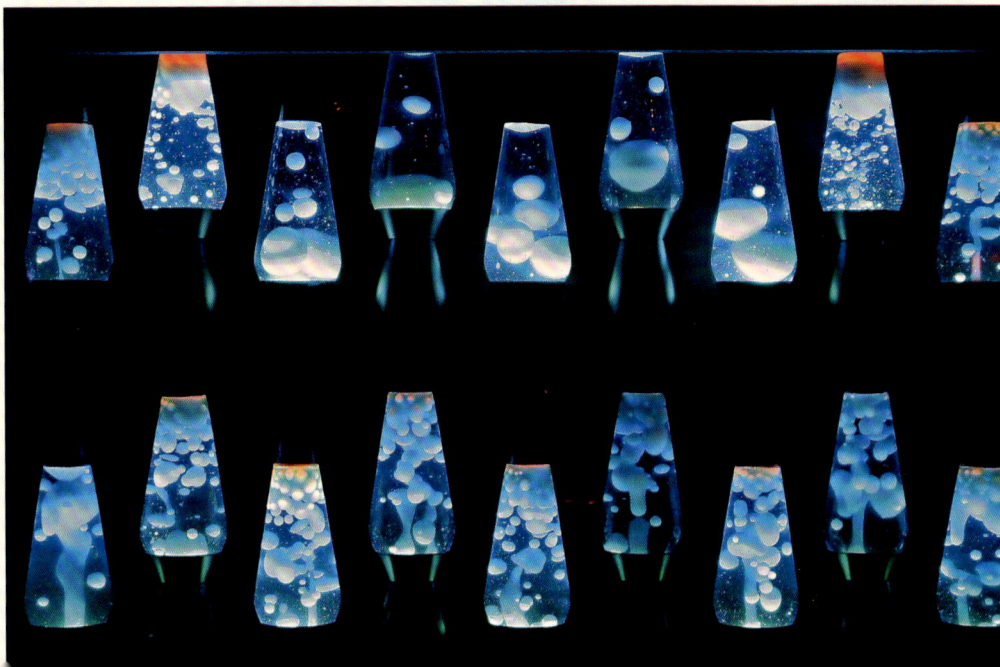

80 Hacking

INNOVATORS ARE REQUIRED TO THINK DIFFERENTLY ABOUT PROBLEMS, and that definition makes them all hackers. The idea being that hacking really means to take old tech and use it in new ways—for good or for ill.

SCIENTIFIC HOOLIGANS

Before hacking was defined, similar activities were described as scientific hooliganism. The first of these hooligans was the magician Nevil Maskelyne (below). In 1903, Guglielmo Marconi demonstrated his radio technology that he had patented. Maskelyne was opposed to Marconi having control of radio systems, which he used in his magic act. Maskelyne used a strong signal to tune into and interrupt Marconi's radio transmission, and tapped out the Morse code for, "rats, rats, rats."

Today we regard hackers as murky cybercriminals (not without foundation), but the basis of hacking culture grew out of a harmless hobby. In 1955, members of MIT's Tech Model Railroad Club were asked not to hack while the network was powered up. This referred to people who decided to do a bit of rewiring or track changes to improve the set— and often blew the fuses in the process. By the 1970s, the idea of hacking had ballooned into an ethos of quickly engineering something useful out of whatever was to hand, and this found its most prolific outlet in the burgeoning field of computing. One of the first groups of computer hackers was the Homebrew Computer Club, which formed in the Silicon Valley area in the 1970s.

Its members toyed with hardware and software, and included Steve Wozniak, who co-founded Apple computers; Dan Werthimer who helped set up SETI, the search for extraterrestrial intelligence; and Adam Osborne who marketed the first portable PC.

Part of the hacker creed is to make all information free, and this has evolved into hactivism, where hackers use their skills to expose people's misdeeds and disrupt wrongful activities.

Hacker culture

The criminal and nefarious hacker who breaks into private computers and steals secrets arose with "Captain Zap," who is reported to have hacked the AT&T phone company in 1981 and changed the company's internal clock, so customers were charged night rates during the day and vice versa. The hacker, real-name Ian Murphy, was tried and convicted for this action. Today's hackers are much more serious fraudsters— known as crackers—who steal billions of dollars from people each year by using various techniques to dupe users into giving away secret information or cause huge damage and disruption to networks using computer viruses.

TRUTH

81 Nanotechnology

NANOTECHNOLOGY IS TECHNOLOGY CARRIED OUT ON THE ATOMIC SCALE. One nanometer is a billionth of a meter—the average sheet of paper is about 100,000 nanometers thick, a pinhead a million nanometers wide.

The concepts behind nanotechnology began in December 1959 when physicist Richard Feynman gave a talk called "There's Plenty of Room at the Bottom" in which he theorized about a process that would allow scientists to manipulate and control individual atoms and molecules. All that was lacking were the tools to do it. More than a decade later, in 1974, Professor Norio Taniguchi coined the term "nanotechnology" to describe his researches into ultra-precision machinery. However, it wasn't until 1981, with the development of the scanning tunneling microscope (STM) that the era of modern nanotechnology began. The STM is a remarkable piece of technology that relies on a phenomenon called quantum tunneling. This arises from the wavelike properties of electrons, and basically allows them to "tunnel" into places ordinary physics says they shouldn't be. The STM makes use of the fact that the probability of finding such tunneling electrons decreases exponentially with distance. The incredibly fine tip (just a single atom wide) of a tungsten needle is positioned a few nanometers from the sample surface and detects electrons tunneling across the gap. Variations in the electron flow are used to build up a picture of the surface. The STM allowed scientists to create images of individual atoms by size and location. In 1989, IBM scientists famously used an STM to manipulate 35 single atoms of xenon on a nickel surface into a pattern spelling out the letters "IBM."

Robert Curl, Harold Kroto, and Richard Smalley discovered in 1985 that carbon can exist in the form of tiny spheres of 60 or 70 atoms, called fullerenes or, less formally, buckyballs. It is possible to enclose metal atoms inside buckyballs and create new organic compounds. A few years later, in 1991, hollow carbon nanotubes, like stretched out buckyballs, were discovered.

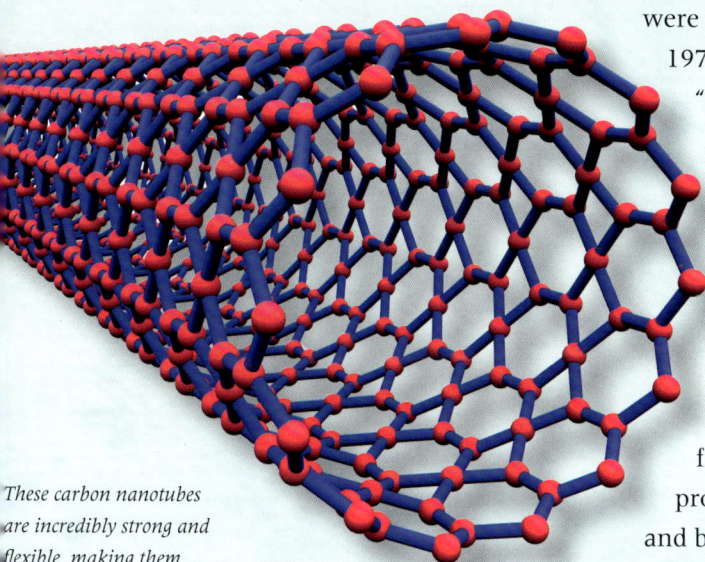

These carbon nanotubes are incredibly strong and flexible, making them potentially useful in many nanotechnological applications.

NANOTECH TODAY

In just a few decades, nanotechnology has begun to play an important role in a number of areas. In medicine, nanotech is in diagnostic biosensors, drug delivery systems, imaging probes, and cancer treatments. In the food industry, nanomaterials have been used to boost production, packaging, and shelf life, to detect food quality, and to protect against food-borne bacteria. Nanomaterials have also been used to explore ways of delivering clean energy in technology such as solar cells and hydrogen fuel cells—and they are in sun creams, too.

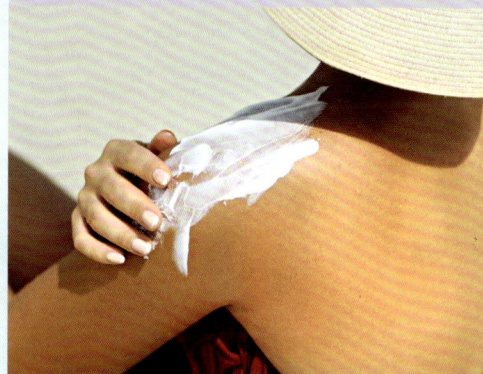

82 3D Printing

IMAGINE A WORLD IN WHICH ORGANS NEEDED FOR LIFE-SAVING SURGERY CAN BE PRINTED FROM HUMAN TISSUE. That is just one of the most eye-catching developments that might be possible with the rise of 3D printing.

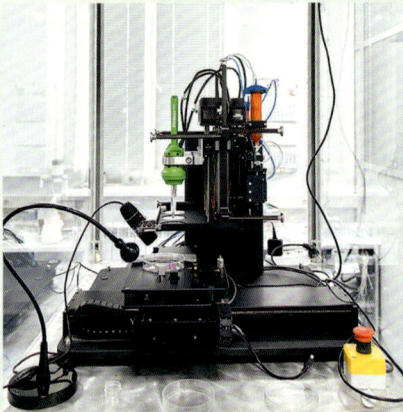

PRINTING LIFE

Bioink is a material containing living cells that is used to print artificial tissue and organs. The idea originated in 2003, when the American Thomas Boland proposed printing using cells supported in a gel that supplied nutrients. Although bioprinting remains experimental, physicians already use it to print blood vessels and skin for reconstructive surgery.

An earlier form of 3D printing, called stereolithography, used a laser to solidify shapes in a tank of liquid resin. As the laser solidified a layer of resin, a platform rose from the base to "push" the object up from the tank in a series of tiny steps. However, In the mid-1970s, engineers began to see inkjet printers in a new way. Instead of spraying spots of ink to shape letters on a page, could the printer produce tiny drops of liquid rubber, plastic, or another material that would solidify, and could the printer continue to build an object, layer by layer, to create a three-dimensional form?

Inkjet technology requires fairly simple technology and has been widely adopted. S. Scott Crump, an American scientist who has been one of the most influential pioneers of 3D printing, promoted it as a method for creating prototypes of industrial machines and for building replacement parts on demand. In the past couple of decades, 3D printing has become a popular hobby, and outlets have appeared on high streets and in malls. For its supporters, this is only the start of the potential for the technology. They envisage a time when 3D printers would make space travel safer and more reliable by allowing astronauts to print replacement parts for their spacecraft. NASA is even experimenting with ways to print food in space. If humans are ever to colonize other worlds, 3D printing will likely be a decisive technology.

Three-dimensional shapes are coded as print files produced from a design created on a computer. This file can be shared so that the object can be created whenever and wherever it is needed.

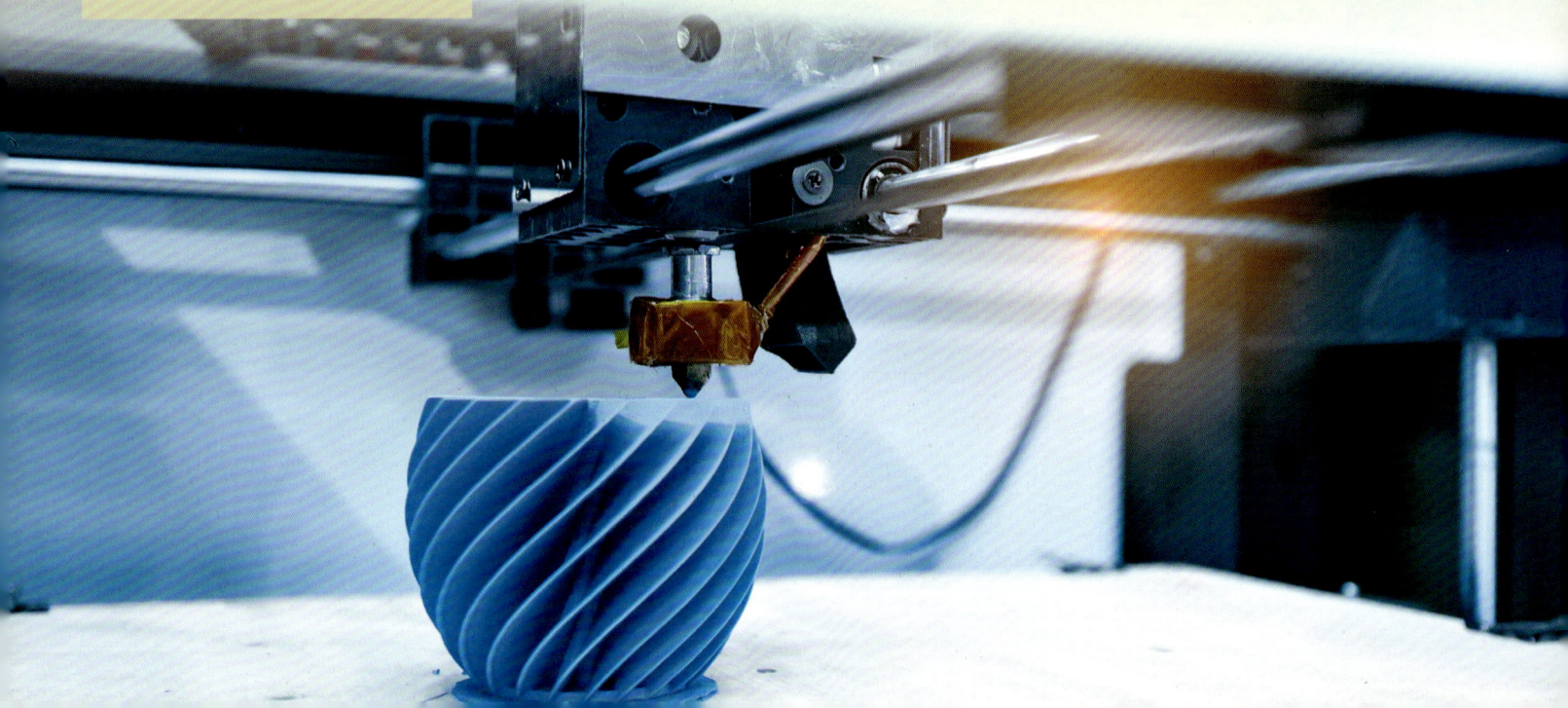

83 GPS

STANDING FOR GLOBAL POSITIONING SYSTEM, GPS WAS INITIALLY DEVELOPED FOR MILITARY USE, but gives us all access to a satellite constellation which pinpoints our exact location just about anywhere on Earth.

The ability to pinpoint location precisely and quickly by GPS has opened up many new areas of technology, from dating apps to the analysis of traffic congestion.

A constellation of a minimum of 24 satellites is needed for global coverage, and the final, 24th satellite was placed in orbit in 1994 (there are now 31 in the system). From just about any point on Earth—unless mountains or buildings get in the way—at least three GPS satellites are overhead at any one time, and can be detected by a unique radio signal. Each signal contains the time it was sent and the location of the satellite. The radio signals travel at the speed of light, but even at that great speed there is a lag between the time sent and the time received. These tiny lags give the precise distance to each satellite. So once received by a GPS-enabled device—a smartphone's map app or a car's GPS or satnav, for example—the GPS device now knows where the satellites are and how far away they are. Those data can be used to calculate the exact position of the device (and you) to within a few feet on the surface of the globe. Devised as a military navigation and targeting technology in the United States, it was always intended to be put to civilian use once up and running. Today, it can be used to track our journeys and locations minute to minute.

GPS satellites are all in a medium-Earth orbit so they arc across the sky at twice the speed of the planet's rotation.

Before GPS, only a trained navigator could calculate their location. In the 1940s, a system known as LORAN used a network of transmitters, which bounced precisely timed signals back and forth. If you knew the distance between transmitters you could calculate your distance from them by the timings of their signals.

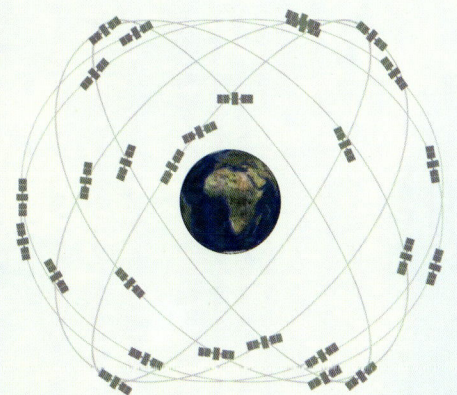

84 VR

FOR SOME USERS, ONE EXCITING OPPORTUNITY THAT ACCOMPANIED THE RISE OF COMPUTERS at the end of the 20th century was the promise of creating alternative worlds, offering experiences that would be impossible in reality.

An early attempt to achieve VR was the Sensorama built by cinematographer Morton Heilig in the early 1960s. The device tried to supplement moving images with an immersive viewing experience that stimulated all the senses (with limited success). Later that same decade, a Harvard physicist named Ivan Sutherland began to explore the idea of creating a head-mounted device, linked to a computer, that would generate a three-dimensional world with which the user could interact. The headset used sensors to track the user's movements so that the computer could update the visual display. Sutherland's 1968 headset was so heavy it had to be hung from the ceiling.

Further research and development has been continuing ever since by organizations such as NASA, the military, and major airlines, for whom VR flight simulators are an effective training tool. VR enables surgeons to practice operations, soldiers to train for battlefield situations, and engineers to navigate their designs. Special forces troops habitually train using VR headsets and laser guns for tasks such as hostage-release missions that would otherwise be impossible to simulate. However, despite obvious applications, VR tech did not initially find a mass market, not even in gaming, which would appear to be highly suitable for it. In the 2010s, more manufacturers began to explore the potential of VR, including Facebook (Oculus Rift), Sony (Morpheus), and Google, which developed Cardboard, a DIY stereoscopic viewer that works with smartphones.

In many ways, VR can be seen as the natural extension of efforts to replicate reality that have been going on for centuries, beginning with the invention of linear perspective in the 1500s. The modern equivalent is a "4D" experience that offers an ever-more immersive viewing.

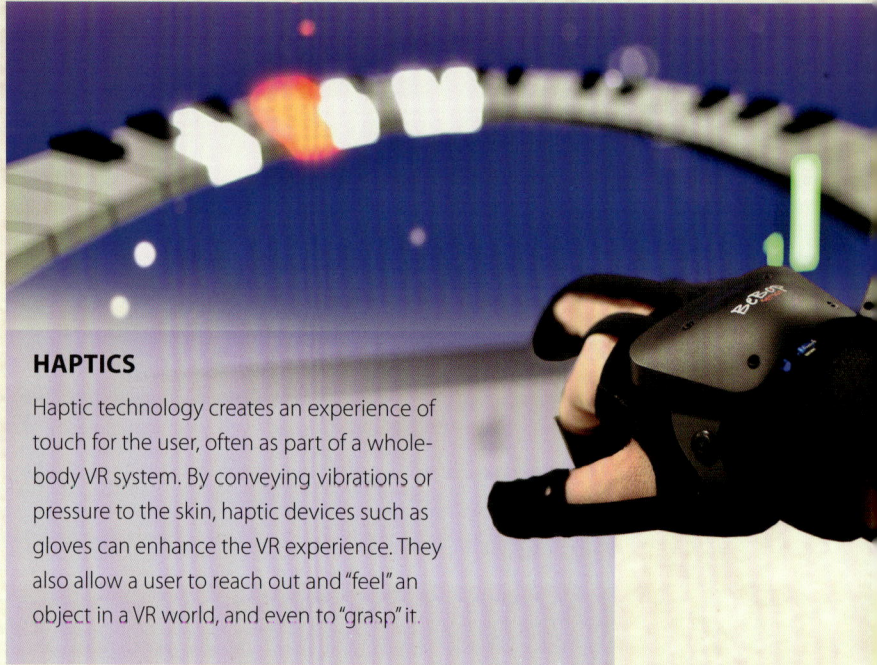

HAPTICS

Haptic technology creates an experience of touch for the user, often as part of a whole-body VR system. By conveying vibrations or pressure to the skin, haptic devices such as gloves can enhance the VR experience. They also allow a user to reach out and "feel" an object in a VR world, and even to "grasp" it.

85 World Wide Web

WE TAKE INFORMATION FOR GRANTED. WE CAN FIND OUT ABOUT ALMOST EVERYTHING FROM ALMOST ANYWHERE. The information is on computers connected by the Internet, and we view it through the World Wide Web.

When Tim Berners-Lee joined the particle physics laboratory at CERN in Geneva as a software engineer in the 1980s he was struck by how difficult it was for scientists to share information. It was easier to ask over coffee than to try to log on to a number of different computers that might not even be using the same programming language.

The solution Berners-Lee came up with involved hypertext—a means of linking stored documents using embedded clickable words. By the late 1980s, millions of computers across the world, including those at CERN, were already linked up via the rapidly growing Internet. In March 1989, in a document called "Information Management: A Proposal," Berners-Lee, with help from his colleague Robert Cailliau, laid out his ideas for linking hypertext documents and the Internet in a system he called the World Wide Web.

By October of 1990, Berners-Lee had set out the three fundamental technologies that still underpin the Web today: HTML, or Hyper Text Markup Language, the Web's markup or formatting language; URL, or Uniform Resource Locator, the unique Internet "address" that identifies each resource on the Web; and HTTP, or Hyper Text Transfer Protocol, which facilitates the retrieval of linked resources from across the Web. By the end of the year, the first Web server was up and running. The first Web pages were simply black and white text—Berners-Lee's personal computer didn't even have a color display. Berners-Lee's real vision lay in his conviction that the World Wide Web would only really take off if it was open to everyone. It was this openness that led to the rapid expansion in accessibility and programming that today means we can use the Internet for research, listening to music, watching TV, or sharing photos of eccentric cats.

In 1993, Tim Berners-Lee got CERN to agree that the Web's underlying code would be available to anyone.

INTERNET OF THINGS

The Internet of Things encompasses everything with a connection to the Internet, which these days means a lot more than just computers. Smartphones, "wearables," such as fitness monitors, and devices that can turn your house heating on or monitor the contents of your fridge, are all part of the Internet of Things, keeping track of our activities minute by minute.

86 Going Wireless

IN THE EARLY DAYS OF THE COMMUNICATION REVOLUTION the focus was on combining computing technology with telephony, and that meant wiring stuff together. Then, in the 1990s, we went wireless.

In the 1990s, communication technology began the transition from analog to digital, where the sound of voices is transmitted as a code of digits or numbers, not as an electronic representation of the sound wave. Soon everything, from radio and television to music and messages, were all combined into a single type of digital data. As a result, all devices are now variants of the same computing technology. Phones, TVs, radios, computers, speakers, microphones … they are all just computers, which share a common language—at least if their programmers allow it.

FROM BLACK HOLES TO HOT SPOTS

The concept of Wi-Fi is straightforward enough. It's a radio link with a gateway to the Internet. As one might expect, in practice it is very much more complicated. The difficulty was that the power of a Wi-Fi signal is set to fill the volume of a house or office and not spill out too far beyond, and the alloted frequencies carry dozens of distinct signals that would diffract and interfere as they bounce around inside buildings. The system that eventually solved this issue and allows Wi-Fi to unpick this jumble of radio was developed by Australian astronomers who aimed originally to use it to find micro black holes (below) in deep space.

Invisible signals

These devices only function if connected without wires; all the connection cables would render them worse than useless. By the same token, this diversification of technology is only possible thanks to wireless tech. There are three main wireless systems: cellular telephony, Bluetooth, and Wi-Fi. The first was developed through the 1970s and 80s, but went fully digital in the 1990s. It uses microwave radio waves to send and receive signals to mobile phone handsets (whether that's voice or email or whatever, is of secondary importance). Instead of connecting directly between handsets like some kind of mass-use walkie-talkie system, the phones connect to a local cell tower, which then carries signals through a wired network to a tower where it makes the final stretch of the journey to the receiver as radio.

Bluetooth was developed in 1992 and named after a Viking king who unified Denmark. Bluetooth is a short-range radio-link that, like its namesake, is a unifier, this time of computing with telephones—and all kinds of other devices. Wi-Fi, a system also formalized in the 1990s, has a router with a wired connection to the Internet. The router sends and receives traffic by radio to and from devices within a 10-meter (30-ft) range.

87 Google

AS THE WORLD WIDE WEB GREW, IT BECAME NECESSARY TO SEARCH ITS CONTENTS TO FIND THE RIGHT PAGES. Early simple search engines were based on keywords, but in 1998 a pair of mathematicians created a new approach.

An early version of the Google home page was very simple—something it retains—because its founders were not very good at coding Web pages with HTML.

For the first three years of the World Wide Web's existence, its contents were listed by hand, with Tim Berners-Lee, the Web's creator, maintaining a record of webpages, which could be accessed on the primordial webserver he hosted. In 1993, primitive catalogs of file names and titles were created by automated systems called webcrawlers. As this name evokes, this software moved steadily through every page of the Web, recording its contents in a searchable database. The first all-text search facility, called WebCrawler, appeared in 1994. However, as the Web grew in size, the time needed to review and update its text contents grew and grew—as did the demand from users to search for something useful and relevant. In this space, several start-up search companies appeared, including Yahoo, Excite, and AltaVista, although at first these search systems were based on a directory of Web sites and pages, and only later began to return results for user's keyword searches based on a page's contents.

As well as searches, Google also provides a wide range of Web services.

Adding quality

Full text search was a big step, but common search terms of the kind entered by non-technical public users of the Web, who were growing in number, turned up thousands of near identical pages that all contained those words. In 1998, Larry Page and Sergey Brin, a pair of mathematicians working at Stanford University, showed a way of ranking pages according to how often they were arrived at through hyperlinks from other pages. The reasoning here was that pages that were visited more often were more useful to us than the ones that were never read. Page and Brin used this piece of math to create the search engine Google, a name based on the number googol, which is 10^{100}. Legend has it the word was misspelled when the Web address was being registered. The Google ranking system listed results by relevance, and this eventually allowed the company to link on-screen advertising to a user's particular needs, a business model which underlies much of today's Web services.

The first Google server was housed in a case made from children's building blocks. Today, the company has at least a million servers.

88 Social Media

THE FIRST DECADE OF THE WEB WAS ABOUT CONNECTING COMPUTERS, SO PEOPLE COULD ACCESS the information they shared. However, the technology has gone on since then to focus on connecting people.

Around the turn of the millennium, the world experienced a short but significant market correction called the Dot-com Bubble. Investors recognized the immense potential of Web and Internet technologies and paid handsomely for shares, only to find that many start-ups were soon exposed as not having a viable means of making money. Beyond the retail offerings, known then as "e-commerce," but now known simply as "shopping," many Web services struggled to make money from customers, or users, who accessed the services for free.

Building platforms

In the end the most successful new services were those that relied on one of the oldest human interests—gossip. The result was social media platforms, such as Facebook, Twitter, and YouTube, where users not only produced the content, such as family photos, pieces of news, and home-made videos, but then came back again and again to have a good look at what other users were sharing. The old Web of home pages and websites was over. Now we had Web 2.0, where people connected to socialize, argue, and snoop. This personal activity was then analyzed to better target advertising, news, and sponsored content. For a free service, they certainly make money. U.S. Social media platforms alone are worth almost half a trillion dollars worldwide. However, the personalized bubble of information that platforms offer users have big downsides. People are cut off from views that dissent from their own. Instead of exposing people to the wealth of human knowledge, Web 2.0 is cutting many off from it.

PEER-TO-PEER SHARING

Part of the success of Web 2.0 is that it relies on users accessing information and services via a single portal, which monitors, guides, and controls what is on offer. The opposing model is peer-to-peer, where material is shared by a distributed network of users. The most successful peer-to-peer model is torrent technology, where a file is downloaded as multiple packets from several—perhaps thousands—of sources.

The influence of social media has meant that every significant experience (and some not so significant) are felt to be worthy of recording and sharing.

89 Streaming Services

THE PROMISE OF THE INTERNET TOOK SOME TIME TO ARRIVE DUE TO RESTRICTIONS ON HOW DATA REACHED USERS. However, today, high quality music, games, and movies can be streamed to every home on demand.

Until the start of the 21st century, most home Internet connections were limited by the bandwidth capacity of a telephone network that had been designed in the 1950s—if not before. The problem was that all the data needed to be sent as a code of audible tones since the lines had been built to carry a human voice. This severely limited the amount of data that could fit into this narrow band of frequencies. Then, the ADSL system increased bandwidth a hundredfold by sending data as if it were inaudible ultrasound.

Today, even faster Internet connections are delivered by a dedicated communications cable, and that makes it quick and easy to download large files that once were beyond the abilities of unstable connections. In addition, high-definition content, be it music or movies, or software running on a remote server, can be streamed through the connection. This is an aspect of cloud computing, where the storage, memory, and processor used for a particular application or program are not in your own device. Instead, the screen and sound outputs from the software are streamed through the Internet to your screen. It is impossible to tell the difference between software running locally or in a data center halfway across the world.

CONTENT COMPRESSION

The likes of Netflix, Spotify, and other streaming services have changed the way we watch TV and movies and listen to music—and how TV, movies, and music are made. This is all possible due to compression technology that reduces the amount of information needed to stream this content. Only the parts of the video frame or musical bar that change are transmitted. All other aspects are left unchanged, and so do not need to be included in the signal. This digital compression is why pictures break up into squares when the signal drops below optimum (above) instead of going fuzzy all over.

90 Bitcoin

A SUCCESSFUL INVENTION IS A DISRUPTIVE ONE THAT SHOWS THE INNOVATIONS OFFERED ARE BETTER ALTERNATIVES TO WHAT CAME BEFORE. THE "CRYPTOCURRENCY" BITCOIN has the potential to be an alternative form of money beyond the control of governments. That would make it one of the most disruptive inventions in history. Time will tell.

The first Bitcoin was spent in 2009 by its creator Satoshi Nakamoto, supposedly a Japanese software developer, although Nakamoto is a pseudonym, and nobody has ever identified their real identity. Nevertheless, Nakomoto's Bitcoin software has definitely had a real-life impact: currently all the world's Bitcoin is worth roughly $250 billion. But how is it worth anything? To understand that, let's look at how bitcoin works.

Mining for value

Bitcoin is a decentralized currency. While the dollar and euro are issued by a central bank with the authority to print more or less, Bitcoin has no controller. Instead, there are rules that set out exactly how many coins there are at any one time. Bitcoins are not minted or printed, but have to be mined before they can be spent. Like a real miner, a bitcoin miner must expend time and energy, only they do this by solving a coded puzzle. The solution is found by a process of mathematical trial and error, and miners use powerful computers to calculate as fast as possible. As the number of mined coins has increased, the system makes the puzzles cracked by miners consistently harder, which ensures that bitcoins are released at close to a regular rate.

Bitcoin miners have to use supercomputers. In 2020, 200 million billion calculations were required to mine each coin. This needs a lot of energy, so some miners have their own power plants—and a correspondingly huge carbon footprint. Bitcoin is not a green technology.

BLOCKCHAIN

Every Bitcoin transaction is recorded on a public ledger called the blockchain. This record ensures that no one can spend their bitcoins more than once. To maintain maximum security, the blockchain is not held on a single computer, but is distributed over a peer-to-peer network made up of the bitcoin-mining computers that host the blockchain as they chip away at the next coin. Every ten minutes or so a new block is added to the chain thanks to the processing work done by the miners. This new block records all the transactions that happened in those few minutes.

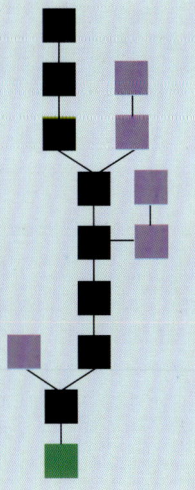

Commodity not currency

Bitcoin is described as a cryptocurrency. This does not mean that its purpose is to hide those who deal in it—although that is an attraction for some. The "crypto" instead refers to the strong encryption used to maintain an absolutely true, unmodifiable record of every transaction. This security creates the trust needed for the currency to work. While people buy and sell services with Bitcoin, almost exclusively through transactions on the Web, economists argue that Bitcoin is not a currency, like the yen or ruble, but a commodity like gold, which will one day run out. It is this scarcity that gives gold its value—and the same is true of Bitcoin. As the number of bitcoins increases, the rate at which new coins are mined drops. The maximum number is 21 million bitcoins, which will happen around 2140.

91 Fitness Trackers

IN 2009, JAMES PARK AND ERIC FRIEDMAN UNLEASHED THE FITBIT TRACKER, one of the first examples of wearable digital fitness technology. Ever since, we've been collecting data about our activity—or lack of it!

In 1965, Dr Yoshiro Hatano of the Kyushu University of Health and Welfare was concerned about the rise in obesity in Japan. In order to promote the health benefits of daily activity, Hatano invented the world's first wearable step-counter, called the *manpo-kei*. The name translates as "10,000-steps meter"—an arbitrary target that has lodged in the brains of the health-conscious ever since. In 1977, Finnish professor Seppo Säynäjäkangas invented the first battery-operated fingertip heart rate monitor as a training aid for cross-country skiers. The company he went on to found launched their first retail product in 1978, a wearable heart rate monitor. In 1984, they introduced the Polar Sport Tester, a heart rate monitor with an integrated computer interface that allowed athletes to view and analyze their training data on a computer for the first time. The Fitbit is a wireless-enabled device that clips onto the user's clothing and uses an internal motion detector to track the wearer's movements and calculate the number of calories burned. It also monitors sleep activity.

Location and activity data from tracking apps helps to create personalized exercise regimes—and is also processed by Web giants to glean facts about users.

92 Ocean Power

THE WORLD'S OCEANS ARE A VAST SOURCE OF CLEAN, RENEWABLE ENERGY, stored in the movement of ocean waves, in tides, and in the temperature gradients between the surface and cooler ocean depths.

The first to patent a system for harnessing power from the ocean waves was French engineer Pierre-Simon Girard in 1799. Although his device was never built, Girard envisaged using wave energy to power saws, mills, and other types of machinery. So far, no one has managed to harness ocean waves on any large scale. Tidal energy is another potential energy source, more reliable and predictable than waves. Whereas wave power is indirectly harnessing the energy of the wind that generates ocean waves, tidal power is exploiting the gravitational energy of the Sun and Moon that causes the rise and fall of the tides.

Tidal power is old tech. The oldest known tide mills, which of course didn't generate electricity, date back to the 7th century. In the same way that wind turns the blades of a windmill, the motion of the tides turns the turbines of a tidal power station. The first tidal power station opened on the Rance Estuary in France in 1966. Still in operation today, the plant's 24 turbines generate 500 million kWh of electricity every year. It remained the largest such installation until the Sihwa Lake Tidal Power Plant opened in South Korea in 2011.

WIND TURBINES

The perturbations of the ocean are matched by the flow of the air, which we call wind. And wind power has been a mainstay of renewable energy for centuries. The first wind-powered electricity generator was built by Scottish engineer James Blyth in 1887, when he constructed a windmill attached to a dynamo to light his cottage. His windmill had a vertical axle and cup-like structures to catch the wind, a design that Blyth believed would prevent the mechanism from being damaged by strong winds. In 1891, Poul la Cour in Denmark developed four-bladed turbines, much like the propeller-like turbines we see towering above the sea today. By 1910, there were hundreds supplying farms and villages.

Ocean thermals

In addition to tides and waves, another form of ocean kinetic energy could be exploited, the energy of ocean currents. Called "underwater windmills," the turbines are placed into an ocean current which turns the turbine's blades to generate electricity. An experimental generator with a capacity of 2 kW was installed 50 meters (164 ft) beneath the sea off the Florida coast in 1976. The world's largest tidal current array stands on the seafloor between the Orkney archipelago and the mainland of Scotland. The 20-meter (66-ft) blades of the four turbines, turning

Oceans are vast, open spaces, where wind blows unhindered for thousands of miles. Wind turbines, built offshore in shallow seas, are proving to be a highly efficient means of generating sustainable low-carbon power.

WAVE POWER

The ability to harness the power of ocean waves is a tantalizing one, but one that so far eludes us. Japanese naval commander Yoshio Masuda developed the oscillating water column, in which air is pushed through a turbine by wave motion. It was used in the 1960s to power navigation buoys. The first "wave farm" opened in Portuguese seas in 2008 using floating, snakelike Pelamis devices (left) made of cylindrical floats linked by hinged joints. The waves cause the device to flex, driving hydraulic pumps that turn generators. The power plant shut down within weeks, and there remains a technological challenge: how to generate electricity in all sea conditions and transfer the power generated to the land.

about once every ten seconds, generate enough power to supply around 2,600 homes. There are plans to extend the facility to 40 turbines.

One of the newest systems proposed is the "wave carpet." Made from a smooth sheet of rubber, the carpet rests near the ocean floor and bends and flexes as waves pass over it. The up-and-down motion drives a piston pump which converts the movement into electricity. The carpet, developed by M. Reza Alam and his team at the University of California, Berkeley, is currently undergoing trials.

Although the concept was first proposed in the 1880s by French engineer Jacques-Arsène d'Arsonval, ocean thermal energy conversion (OTEC) is one of the more experimental types of ocean power tech. It exploits the temperature differential between the warm surface and the cold depths of the ocean, a temperature gradient that can be as much as 50°C in some places. OTEC technology needs a minimum temperature difference of 20°C to achieve significant energy yields. The warm surface water is used to evaporate a volatile fluid, producing a vapor that drives a turbine-generator, producing electricity. The fluid is cooled using the cold water from the depths and the cycle begins again.

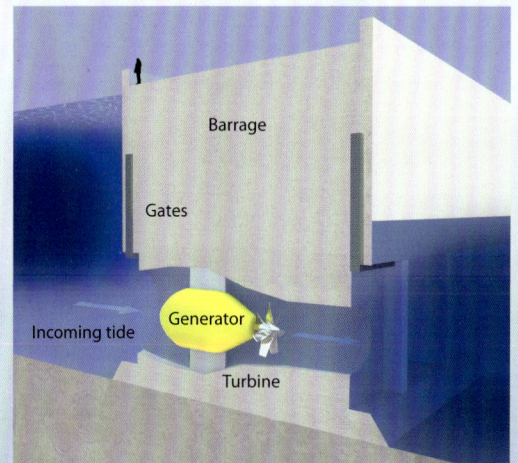

Tidal power works by forcing the incoming and outgoing tidal currents to flow through tunnels in a dam-like barrage and drive turbines inside.

93 Genetic Modification

IN SOME RESPECTS, GENETIC MODIFICATIONS HAVE BEEN PRACTICED FOR MILLENNIA as plant and animal breeders deliberately selected the traits they wanted to perpetuate in future generations. Genetic engineering is a more powerful means of doing this.

In 1971, Kathleen Danna and Daniel Nathans at Johns Hopkins University published their research into bacterial proteins called restriction enzymes, first discovered in the 1950s. These enzymes are produced by bacteria to protect themselves from viruses by chopping up invading viral genes. They are like little DNA scissors.

The first breakthrough in modern genetic technology came in 1973, when Herbert Boyer and Stanley Cohen successfully used restriction enzymes to transfer a gene between different strains of bacteria. A year later, Rudolf Jaenisch and Beatrice Mintz introduced foreign DNA into a mouse embryo for the first time. By 1982, genetically modified *E. coli* bacteria were being used to produce human insulin, the first genetically engineered human medication.

A recent genetic technology promises to be groundbreaking. CRISPR is an easy to use, efficient gene-editing tool, first demonstrated in 2012 by Jennifer Doudna at the University of California, Berkeley, and her collaborator, French scientist Emmanuelle Charpentier. CRISPR is based on the immune system in bacteria that uses enzymes to cut away viral DNA and store a copy of it. Researchers figured out a way of using this copying system to introduce new genes into any cell, making it a more finely tuned tool than restriction enzymes. CRISPR technology is being employed to search for cures for cancer and other diseases. In agriculture, the genomes of plants can be altered to make crops more pest-resistant, better able to withstand drought, or able to yield more fruit. The technology was used to edit some of the vaccines against COVID-19 that were based on messenger RNA.

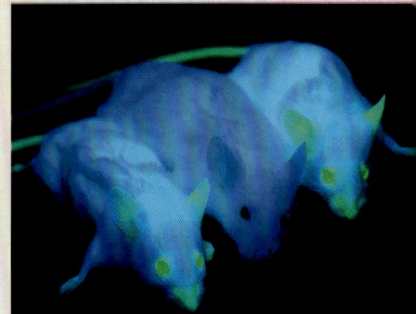

The CRISPR system used by bacteria to detect and repulse attacks by viruses forms the basis of a powerful gene-editing tool. CAS stands for CRISPR-associated proteins—these are the enzymes that do the work.

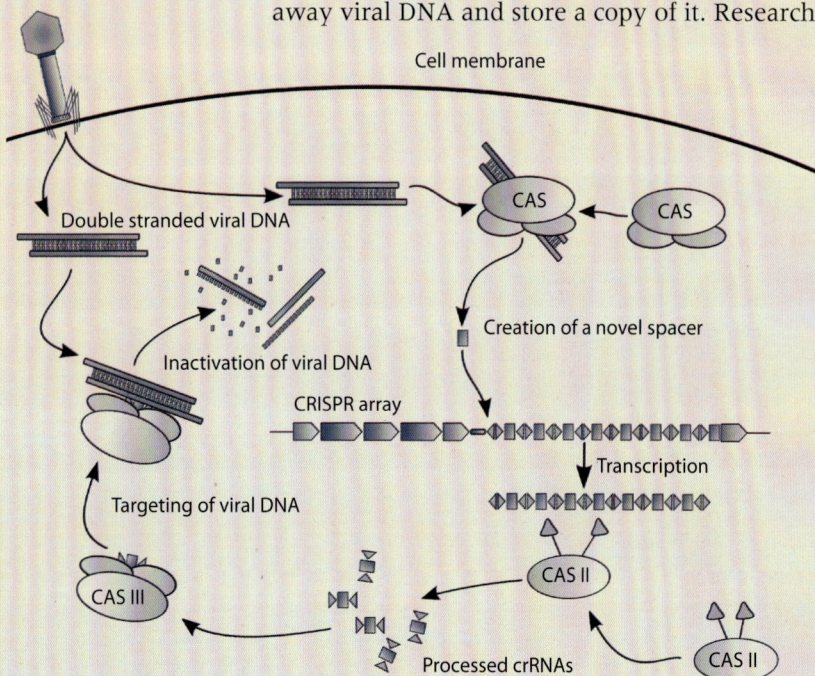

GMOs

A GMO is a genetically modified organism, a living thing—plant, animal, or microorganism—that has had its genetic code altered in some way. Producing a GMO is a much more precisely targeted activity than the gene mixing that occurs in selective breeding. In most instances, GMOs are altered by the addition of DNA from another organism, and are sometimes referred to as "transgenic" organisms. For example, the mice shown above glow in the dark thanks to proteins from a jellyfish, and genes for spider silk have been added to goat DNA to breed animals that produce silk in their milk. This makes a lightweight, ultra-strong material with a range of industrial and medical applications.

Cell membrane

Double stranded viral DNA

CAS ← CAS

Creation of a novel spacer

Inactivation of viral DNA

CRISPR array

Transcription

Targeting of viral DNA

CAS III

CAS II

Processed crRNAs

CAS II

94 Robotics

IT IS BY NO MEANS CLEAR WHAT A ROBOT ACTUALLY IS, BUT IN GENERAL IT HAS CERTAIN ATTRIBUTES, including the ability to perform tasks autonomously and to sense and manipulate its environment.

Valkyrie, developed by NASA in 2013, is a robust, rugged, entirely electric humanoid robot. Originally designed for work in dangerous disaster zones, one day robots like this could form the advance party on missions to the Moon or Mars, preparing the way for human explorers.

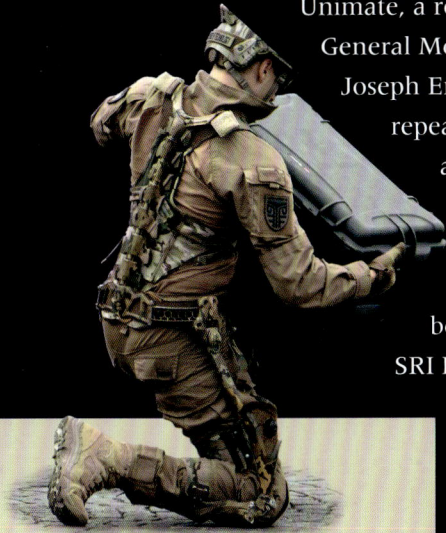

Unimate, a robot arm used for welding car bodies, was introduced to the General Motors production line in 1961. Designed by George Devol and Joseph Engelberger, it was the first industrial robot, and operated by repeating the exact same movements over and over. Today's factories are a forest of robotic technology, performing tasks quickly and precisely that human workers might find tedious and even dangerous.

At around the same time that robotics was beginning to make an impact on manufacturing, SRI International in California developed Shakey. Equipped with a camera and collision detectors, Shakey was built to move autonomously, rolling around in any location, but it was awkward and slow.

The first robot that could walk was the Waseda-Hitachi Legs 11, first demonstrated in 1985. With its hydraulic-powered legs and on-board computer guidance system, it could manage a less than speedy 13 seconds per step on a flat surface. But it was also able to turn left and right, which was a major step forward (or sideways). By 2019, Boston Dynamics' Atlas robot had literally come on in leaps and bounds. The 1.8-meter-tall (6-ft) robot can jump across gaps, leap over obstacles, do back flips, keep its balance when pushed, and jog over a snowy field.

Advances in technology, including computer processing power and precision machining, have allowed the development of more and more sophisticated robots. Robots have come to play an increasingly important role in space exploration, accomplishing missions that would be virtually impossible for a human crew. The addition of finger joints created dexterous robots that could delicately handle objects. In 1985, Dr Yik San Kwoh invented a robot-software interface that allowed the first robot-aided surgery to be performed.

EXOSKELETONS

A powered exoskeleton is a wearable machine that augments the strength and speed of a human operator. The first, called Hardiman, was produced by General Electric in 1969. Unfortunately, only one arm worked as planned. The machine juddered violently when the full frame was powered up, and it was never used with a person inside. More successful prototypes are being developed for use by infantry soldiers and for people with mobility issues who cannot walk unaided.

95 Drones

UNCREWED, REMOTE-CONTROLLED OR AUTONOMOUS FLYING MACHINES HAVE LONG BEEN OF INTEREST TO MILITARY INNOVATORS. But, today, drone technology is finding more and more civilian applications.

Smile, you are on camera. Camera drones are an increasingly common aspect of air traffic. Most countries require operators of drones above all but the smallest toy sizes to be accredited, and ban flights close to buildings.

The advantage of drones was obvious from the early days of aviation, especially when it meant keeping a human operator safe from harm during warfare. An early version was the Hewitt-Sperry Automatic Airplane, which had a limited use at the end of World War I as an aerial torpedo. The drone flew in a straight line and was programmed to drop a bomb after traveling a preset distance. The pilot-less aircraft is a precursor of the cruise missile, as is the V-1 flying bomb, or "doodlebug." This World War II flying bomb, developed in Nazi Germany, flew in a straight line for a predetermined time and then dropped from the sky.

However, UAVs (unmanned aerial vehicles) promised so much more than a crude way to deliver weapons. The next generation of drones were based on small, fixed-wing aircraft, equipped with surveillance equipment. There was no limit to size, with passenger jet test flights being completed with no one aboard, all under radio control from the ground. However, that kind of uplink limited range. Again, it was military demands that led to an upgrade, where drones capable of many hours of autonomous flight were controlled via satellite links by pilots sitting in virtual cockpits on the ground half the world away. A grounded crew could fly the drone over any point on Earth—and get a live, bird's-eye view of the scene below, and a big draw for civilian drones has been the ability

COMBAT DRONES

Modern militaries rely heavily on armed drones to attack targets. Not only does this technology keep personnel safe, it gives commanders with enough drones the reach to attack targets anywhere on Earth just about any time, which will surely be an aspect of future wars. One of the first of these drones was the Predator, which entered service in the 1990s and could stay aloft for 24 hours at a time. The faster and more powerful Reaper entered service in 2007. It could fly autonomously, using its own sensors to navigate along a preset course, without the need for a human pilot back at base. The Reaper has since been joined by the Avenger (below) which is powered by a jet engine rather than a propeller like older models.

The EHang 184 passenger drone can carry a single occupant on a 20-minute flight at 130 km/h (81 mph). The human on board has no flight controls and just chooses the destination.

to take aerial footage cheaply and easily, where once a helicopter was needed. However, fixed-wing drones are unsuited to this task and so smaller drones use multiple rotors. These engines are far more power-hungry than those of a fixed-wing aircraft, but offer much more agility in the air. Improvements in batteries and electric motors have meant that rotor drones have grown in size. In 2014, the Chinese-made EHang 184 UAV was unveiled, and has since been upgraded. Designed as a passenger drone, it uses eight rotors, each with a separate power supply to reduce the dangers of engine failure.

Using drones for delivery is one application that is still uncertain. It is nearly always cheaper to send packages by other means. However, if the package is urgent enough and the destination remote enough, a drone may well be the answer.

96 Near-Field Technology

CASH HAS BEEN AROUND FOR MORE THAN 2,700 YEARS, BUT ITS DAYS MIGHT BE NUMBERED. CONTACTLESS PAYMENTS use a tiny, short-range radio link to perform a transaction. Why bother with coins and bills?

BIOMETRICS

Passports and other forms of ID increasingly rely on biometric information, most obviously a photograph of the owner. This information is digitized and shared by scanning the passport. This is then checked against a scan of your face. If the shapes match up, specifically the distances between certain facial features, then you're in (or perhaps not!). Additionally, in many passports the biometric data is stored on an RFID chip within the cover which transfers the information via a near-field receiver.

By 2014 paying small amounts of money without ever needing cash became widespread. Simply tapping a bank card on the card reader was enough. Even if you'd forgotten your wallet, all the necessary details could be sorted in your phone, making it a virtual wallet. The same kind of contactless technology as used in the card allows the smartphone to transmit the necessary financial details.

That innovation is called near-field technology. Obviously a bank card has no battery, so how does it work? The card has a coiled antenna, or tag, inside, which becomes electrified when a powerful radio source is close enough—a few inches at most. Once electrified, the tag sends out a unique radio signal, which is picked up by the card reader. That is enough to prove the identity and bank details of the payer, at least for small sums. Smartphone payment apps work in the same way, except, to boost security, the phone requires a signal back from the reader.

Contactless payments are being rolled out worldwide.

97 Autonomous Vehicles

THE FIRST CARS WERE KNOWN AS "HORSELESS CARRIAGES," SO IN HINDSIGHT IT WAS ONLY A MATTER OF TIME BEFORE INVENTORS came up with "driverless cars." These high-tech vehicles have been widely tested on roads since 2015, but have some way to go before they revolutionize travel.

People dreamed of creating self-driving cars as long ago as the 1920s, but only since the advent of computers has this dream become close to reality. In 2011, Nevada became the first U.S. state to allow these cars on its—mostly fairly quiet—public roads. By 2015, testing was being carried out in most states and across the world as motor manufacturers recognized the potential for automated vehicles and teamed up with technology companies.

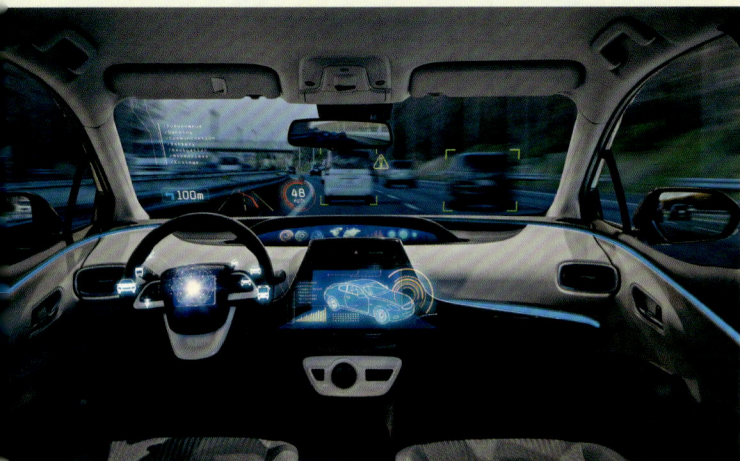

An autonomous car is able to track its surroundings with great precision. It can identify objects and measure their relative motion.

Evolutionary design

Work on a computer-controlled, or autonomous, vehicle began at Carnegie Mellon University in Pittsburgh in 1984. Their "Navlab" vehicles were converted vans—they needed the space inside for all the computer hardware—and they relied heavily on GPS to tell them where they were. Safety concerns meant these vehicles had a top speed of 32 km/h (20 mph). The early collaborations between auto and tech sectors created vehicles covered in outlandish and unsightly detectors and with barely any room for passengers. However, these early testbeds steadily evolved into something more like the sleek, luxury road car that would appeal to customers willing to pay the premium for a self-driving car. As yet, no such vehicles are available for purchase. Instead, some models have an autopilot function, which manages road positioning, maintains safe distances from surrounding vehicles, and brakes automatically in emergencies. However, a human driver must be in control at all times.

An autonomous car has no need for the paraphenalia of driving. Already car manufacturers are beginning to wonder what travelers of the future might want in an automobile.

THE SHARING ECONOMY

Why would you want to own a self-driving car? Indeed, would the autonomous vehicle want to be owned? One of the great hopes of the autonomous era is that today's ride-hailing and ride-sharing apps will become the normal mode of road travel, especially in congested cities. The autonomous cars will patrol the streets looking for customers and set prices according to demand. Potentially the vehicles would become emancipated and be in charge of their own maintenance.

Detect and evade

Self-driving vehicles still use GPS to place themselves on the map, so they can navigate routes and change lanes for turns and intersections. However, GPS is not able to show the car what is around it, such as other road users, pedestrians, road signs, and traffic signals. For that the cars use a combination of live video, a laser-based radar known as lidar, and ultrasonic sensors. The video is provided by cameras pointing front and back, and side to side. AI will identify what can be seen in each view, differentiating between a car, a cyclist, a pedestrian, or a cat or road sign. This job is aided and augmented by the lidar, which gives information about the distance to each object and their speeds and direction. Ultrasonic sensors are already in wide use as parking sensors where they are used to give proximity warnings as you edge closer to obstacles. This same tech is used in self-driving cars to monitor how close the other cars are as they speed past, and tells the car to fine tune its road positioning.

The hardware is now in place, but the software needs more time to be able to replicate the abilities of a human driver navigating along crowded streets and speeding highways. During tests, the vehicles have often crashed because they cannot recognize obstacles, and their video processors rely on clear markings to show where the road is—and these are not always there. Nevertheless, experts predict that by 2040 up to 75 percent of all road vehicles will be autonomous.

CONGESTION TRACKERS

A likely development in autonomous vehicles will be their ability to communicate with each other in ways that are far beyond most human drivers. This will boost safety and create a more smoothly running traffic system. In fact, human drivers might be too unpredictable to be allowed out in public in future! Already the data collected from our cars is being put to use, not least in congestion tracking. This collects location data from the phones of people traveling in cars and uses that to plot where roads are likely to be jammed. Other users approaching the area are diverted along less busy routes—and the jam clears faster.

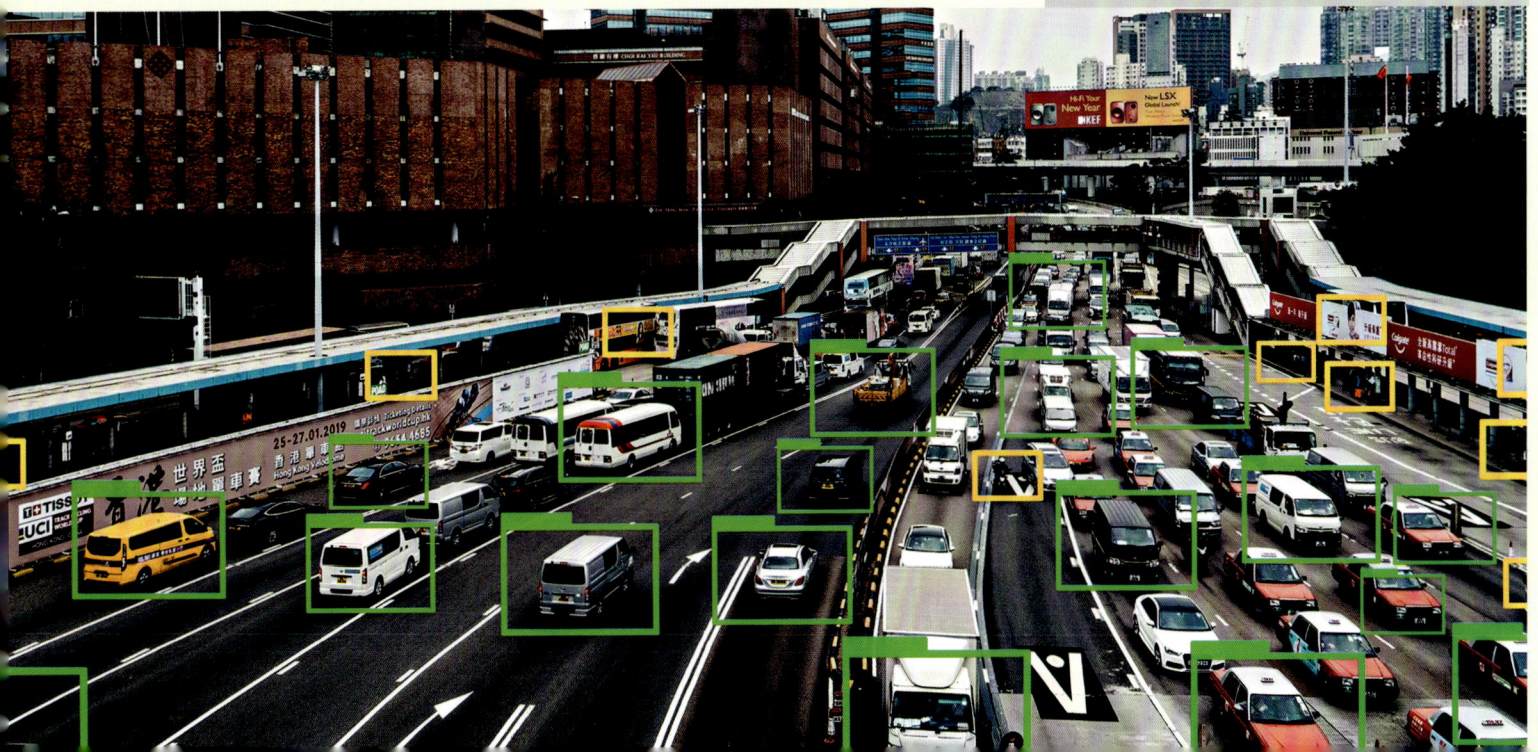

98 Vertical Farms

THE IDEA OF GROWING PLANTS IN THE AIR HAS BEEN AROUND SINCE ANCIENT TIMES. Although there are challenges in setting up a vertical farm, it seems certain that the technology will become a crucial tool in food production.

It is likely that the famed Hanging Gardens of Babylon had terraces rising up a large ziggurat or pyramid. The modern idea of vertical farming is to do the same inside high-rise buildings. This idea is credited to the Columbia University professor Dickson Despommier, who in 1999 set students a task of investigating how much food could be grown on the rooftops of New York City. He later calculated that a 30-story farm could produce enough food to feed 50,000 people. Animals such as pigs on the lower floors would live on vegetable matter from the crops growing above, while the animal waste would provide additional nutrients for the plants.

Growing plants indoors, in vertical layers without soil, promises to increase yields far above what was possible for traditional methods of farming. There are difficulties, however, particularly in the cost of providing growing lights and water to irrigate plants. However, various smaller-scale vertical farming operations have proved highly productive. They use technologies such as hydroponics, in which plants are grown without soil. Instead, plants grow in a liquid full of nutrients, with gravel to support the roots. In aquaponics, a hydroponic system is linked to tanks containing fish, whose waste provides nutrient-rich water for the plants, which in turn purify the water to be returned to the fish tanks. A third technology, known as aeroponics, was originally designed as part of research into growing crops on spacecraft. It grows plants in a soil-less environment by misting the air with nutrient-rich moisture spray.

Hydroponics reduces the amount of water needed to produce fast-growing salad crops, such as lettuce.

99 Cultured Meat

IMAGINE ENJOYING A STEAK THAT WAS ONE HUNDRED PERCENT BEEF BUT HAD NEVER BEEN ANYWHERE NEAR A COW. Unlike traditional meat obtained from an animal, cultured meat is grown from animal cells in the laboratory.

There are sound environmental reasons for developing cultured meat production. More than half of the world's agricultural land is used for beef production, but beef makes up less than 2 percent of the calories consumed by the global population. As far back as 1894, French chemist Pierre-Eugène-Marcellin Berthelot predicted that humans would be dining on meat grown in a lab rather than from once-living animals. The earliest successful cultivation of muscle fibers in the laboratory was achieved by American pathologist Russell Ross in 1971, who grew cells obtained from a guinea pig. However, research into cultured meat production only really took off in the 21st century. The process begins by growing animal cells in the laboratory to establish a culture, or cell line. Cell lines can be based on tissue cells, muscle or fat cells, or stem cells. Once the cell line has been established, a sample is placed in a bioreactor, a vat of culture medium where the cells grow rapidly. Conventional meat derives its flavor and texture from a variety of cell types, including fat as well as muscle, which is still tricky to replicate.

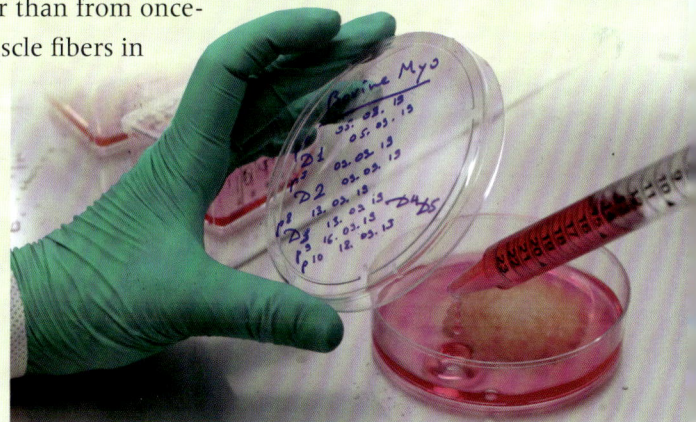

A growing piece of meat is fed with nutrients in an experimental food lab.

NASA became interested in growing meat from turkey cells in 2001 in hopes of providing a food source for astronauts on long voyages. They also funded research into fish protein. Cultured goldfish muscle tissue was dipped in oil, breaded, and deep-fried. Although everyone agreed it looked and smelled good, no one could actually taste the fillet as eating experimental products is forbidden by U.S. law!

The first artificial meat hamburger tasted good but did not look authentic.

The first laboratory cultured beefburger, produced from cow stem cells, was grown from around 20,000 thin strips of muscle tissue. In August 2013, it was cooked and eaten at a press conference in London. At an estimated production cost of $300,000, it was probably the world's most expensive burger. In September 2019, muscle tissue was grown aboard the International Space Station. Bovine cells harvested on Earth were taken into orbit and assembled with a 3D bioprinter, mimicking natural tissue regeneration processes. Although it is space-age tech, artificial meat might soon be widespread down on Earth.

100 AI

IT IS DIFFICULT TO DEFINE INTELLIGENCE EVEN BEFORE INTRODUCING THE TERM "ARTIFICIAL." Artificial intelligence (AI) has been focused on human abilities such as language and problem-solving. However, recent breakthroughs suggest that computer intelligence will be completely different to our own.

SMART TECH

Well over a quarter of U.S. homes now have more than three smart home devices, but there's no easy definition of what "smart" devices are. Broadly speaking, they might be thought of as gadgets we can interact with on some level. The term "smart" originally meant "Self-Monitoring, Analysis, and Reporting Technology," but came to embrace the notion of inanimate objects, from cars to refrigerators and smartphones, smart speakers, and smart TVs, all reacting to what we do and talking back to us.

In 1950, Alan Turing addressed the problem of determining whether or not a machine can be said to be intelligent. The test he devised still illuminates the subject today. Turing proposed that a human questioner interrogates two unseen respondents by means of written questions. One of the respondents is human, the other a computer, and the challenge is to see if the questioner is able to determine which is which. For many years it seemed that although computer chatbots could be understood, they still lacked an authenticity to their language that came naturally to a human. But that all changed in 2022 with the launch of ChatGPT, Bard, and other natural language models, which were able to generate text that was indistinguishable in structure to that of a human author. What was different? How were they created? *Bard was created by Google AI using a process called machine learning. Machine learning is a type of artificial intelligence that allows computers to learn without being explicitly programmed. In the case of Bard, the computer was trained on a massive dataset of text and code. This dataset included books, articles, code, and other forms of text. The computer was then able to use this data to learn how to generate text, translate languages, write different kinds of creative content, and answer your questions in an informative way.* That last passage was written by Bard itself.

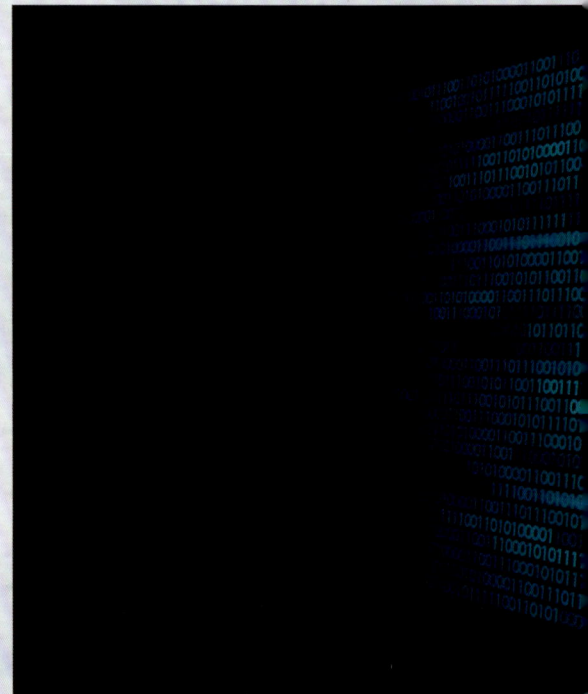

In common with similar systems, Bard has trained itself to be very good, almost flawless, at guessing what word comes next in a sentence. That is the basis of reading and writing (or talking). A major purpose of Bard and ChatGPT is an extension of web search. We ask them to find out something for us, and instead of curating a set

FACTS AND FAKES

Inherent dangers are posed by generative AIs. Firstly, chatbots can talk the talk, but are often saying nonsense. AI-generated text can be full of factual errors because the AI is as yet unable to discern the veracity of the different sources it is consulting. Perfect language lends plausability, but this is not a person speaking, it's an AI that has no understanding of what it says. So watch out. The second concern is deepfakes, which are fictional pieces of audio and video made using AI that are impossible to tell apart from reportage of actual events. That threatens public and private life in innumerable ways. What can we trust if we can't trust our own eyes and ears?

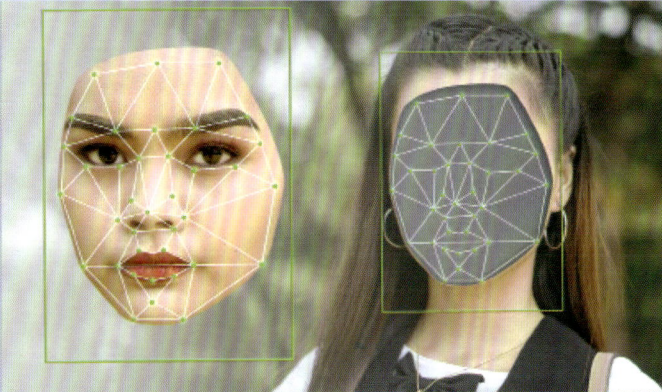

of primary sources relating to our query, the chatbot simply writes out the answer.

Still narrow

In terms of intelligence, even the most voluble chatbots are narrow AIs. Narrow AI systems perform one particular task, such as facial recognition or running a self-driving car. These systems typically rely on machine learning, and eventually they will perform better than a human can in terms of speed and accuracy. General AI applies to machines that are capable of learning in a similar way to humans (that have EI, evolved intelligence), in that the AI knows what it does not know—and needs to find out—to perform a task. Despite the apparent abilities of ChatGPT, a general AI still seems some way off.

General AI would be a digital analog of human consciousness, and we are unsure where consciousness comes from. One idea is that consciousness emerges from the many interconnected systems in our brain. Given a computer powerful enough, would a consciousness emerge? If it did, how would we even know?

Machine learning uses a processor called a neural network, which trains, or programs, itself to perform its task by trial and error as more data is fed into it. The neural network learns by identifying which routes through its many layers correspond to particular inputs. Once that training is complete, it can identify targets with great accuracy.

Building intelligence

Google's AI division, DeepMind, used its AlphaGo AI (a game-playing system) to train its own playing partner, called AlphaZero—and taught it to be an even better player. DeepMind's AIs are focused on very complicated puzzles, but other AIs are generative, in that they produce content. Chatbots generate text, while other generative AIs make pictures, videos (see box, above), music, and computer code. All AIs can pass on what they have learned more easily than humans can. The ability for these generative AIs to train each other and share their programming could create a novel kind of intelligence. While it is overmatched by human intelligence in some ways, these AIs, if left unchecked, could develop at rates well beyond the speed of human learning. This in itself is a reason to slow down AI development, the experts say. Otherwise we risk the uniquely new position of creating technology that understands us better than we understand it.

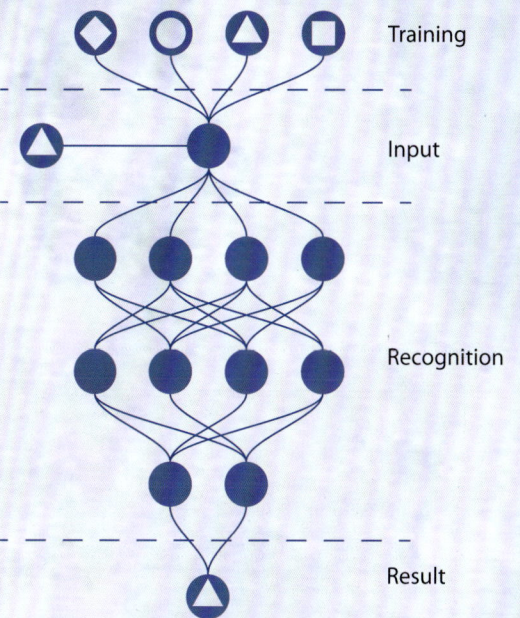

Training
Input
Recognition
Result

101

Technology: the basics

So where does all this innovation and invention come from? Technology is the application of scientific knowledge. It's time to take a tour of some basic—and some not so basic—concepts that lie at the heart of high tech.

Using energy

The Laws of Thermodynamics Sometimes you have to look quite hard for it, but all technology is using a source of energy and transforming that energy into something useful: a motion, a product, or a computer output. That is technology at work, and it is telling that physicists describe the transformation of energy from one form to another simply as "work." Work can include all kinds of energy (although technology makes the most use of heat and electricity), and all forms of energy are bound by a set of universal rules called the laws of thermodynamics. Thermodynamics is the area of physics that covers the behavior of energy. The term thermodynamics means something like "changes of heat," and that is where the understanding of energy began. But the subject throws light on all kinds of energy, as shown opposite.

The three (or perhaps four) fundamental laws were developed throughout the 19th century. Before then, heat and light were assumed to be materials with weird properties, but, bit by bit, it was shown that they were part of a broader pantheon of energies, all capable of doing work. This is the first law: energy cannot be created or destroyed but it can be transformed into different forms, such as motion, heat, light, or sound.

The second law says that a closed system will increase in entropy over time. Entropy is a complicated concept, one best understood as a statistical feature of energy transfer. It means energy is always becoming more dispersed and diffuse, so heat moves toward cold, and ordered structures decay into disorder.

The third law says that the lowest possible temperature is absolute zero (0 Kelvin, or −273.15°C; −459.67°F). It is impossible to reach this temperature because it would need an infinitely sized cooler, an infinite amount of time, and infinite energy to do so. There is also another law of thermodynamics which says that two connected systems will enter thermal equilibrium. That means a hot and cold space will eventually both become the same temperature. Although this rule was well understood long before the three formal laws were set out, it was the last one to be included as a law, but was put first in the list as the "zeroth law."

Types of Energy

KINETIC ENERGY
Many forms of energy can be understood simply in terms of motion, and therefore are all types of kinetic energy. Kinetic is from the Greek for "moving" (we get the word cinema meaning "movies" from this, too). The motion of kinetic energies is found at all scales, from the smallest quantum mechanics to the motion of galaxies.

POTENTIAL ENERGY
Energy can take a "frozen" form, where it fills a substance but is not available for use until it is released as another form. Machines, engines, and electricity generators generally use a source of potential energy, mostly a chemical or nuclear fuel.

MECHANICAL
The energy of motion on the largest scale, relating to objects traveling through space or rotating around a location. This form of energy is transferred from one object to another by collisions and contact.

CHEMICAL
Energy is needed to hold atoms together as molecules. During a chemical reaction, the bonds break, releasing energy which is then used to make new bonds. A fuel or explosive is a chemical that releases a lot more energy in a reaction than it takes in.

ELECTRICAL
The energy that is being transferred as electrons or other charged particles and moves through substances as currents. This energy is based on the fundamental property of electromagnetic charge.

NUCLEAR
The energy stored at the heart of atoms in the protons and neutrons that make up the nucleus. The energy is released when the nucleus is disrupted and changed in a fission or fusion reaction and radioactive decay.

THERMAL
The energy of heat, which is based on the kinetic energy of atoms and other fundamental particles. These are never still, but vibrate or bounce around at random. As this motion increases, the material gets hotter. Temperature is a measure of average thermal energy in a substance.

GRAVITATIONAL
It takes work to lift an object, and, once elevated, that added energy is transformed into gravitational potential energy. When the object is allowed to fall, the object's potential is transformed into motion.

RADIANT
Energy carried by light and waves of other forms of electromagnetic radiation, such as gamma rays and infrared. The electromagnetic radiation is a vibration in the electrical and magnetic fields that permeate space.

ELASTIC
Substances that can be deformed temporarily by a force, but spring back to the original shape are elastic. (Plastic materials maintain the deformed shape.) The energy added by the deforming force is stored in the material and then released as a restoring force.

SOUND
Also known as acoustic energy, this is carried by waves of pressure that travel through a medium, such as air or water. A sound wave is made up of the successive compression and rarefaction (spread) of the medium.

Biotech

What is DNA? The biotech industry seeks to modify life-forms or deploy evolved chemical processes in artificial ways. To do that it must work with the biological information system that is DNA. DNA is a complex polymer with the full name deoxyribonucleic acid. A DNA molecule takes the form of a double helix, or twisted ladder. The sides are made by chains of ribose sugars, while the rungs are constructed from nucleic acids, or bases. There are four bases, two are pyrimidines, and two are larger forms called purines. Every rung is composed of a pair of bases, one pyrmidine and one purine.

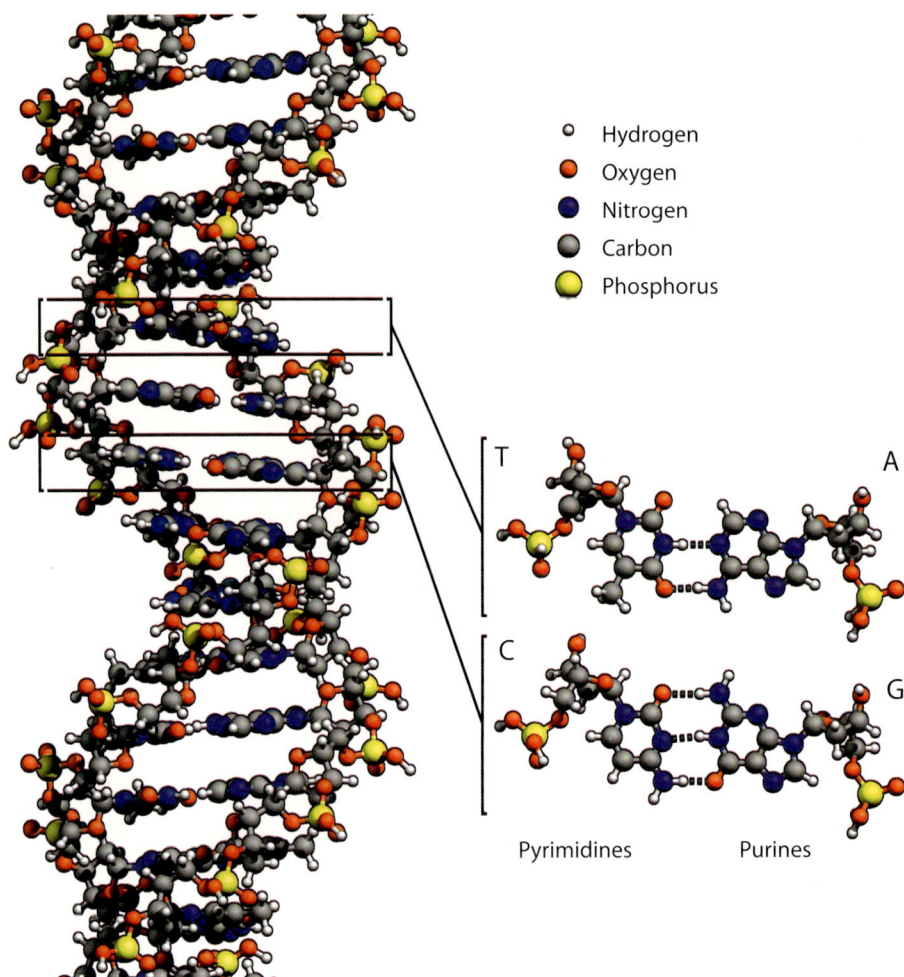

○ Hydrogen
● Oxygen
● Nitrogen
● Carbon
● Phosphorus

T | A

C | G

Pyrimidines | Purines

Differences between DNA and RNA

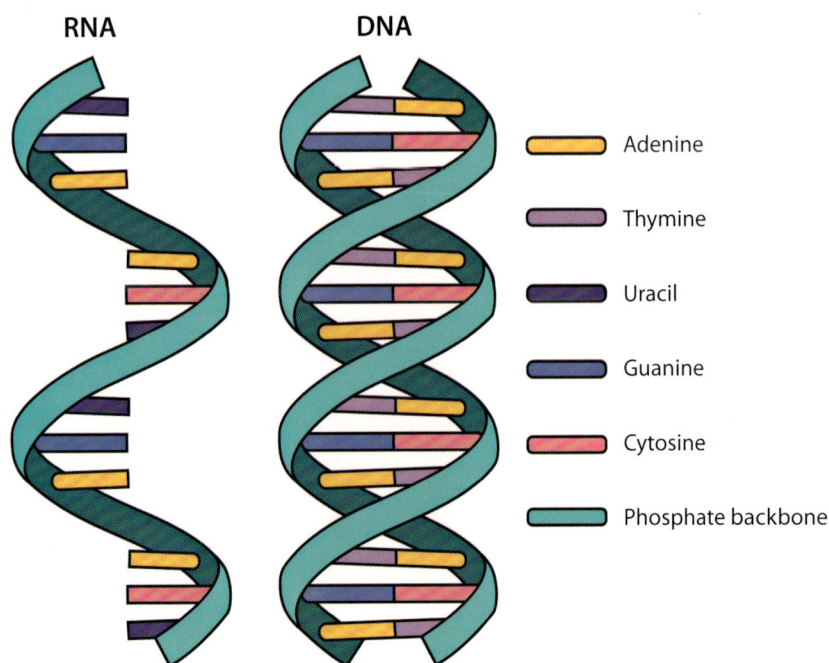

RNA   DNA

■ Adenine

■ Thymine

■ Uracil

■ Guanine

■ Cytosine

■ Phosphate backbone

Four-letter code In DNA the bases are adenine, a purine, and its pyrimidine partner thymine; and the second purine guanine and its partner cytosine. Together the bases of DNA make up a code of four characters, or four letters: ATGC. These letters are arranged in a very specific order within a molecule, and several hundred together are used to encode one gene. To read the code, the DNA helix is unzipped into two strands and a copy is made of the coding strand using RNA, or ribonucleic acid. RNA is a robust cousin of DNA built with the same bases as DNA except the thymine is replaced by uracil.

The genetic codebook The genetic code in DNA carries instructions on how to make proteins. These proteins are the building blocks and machinery of a living cell. Every protein has a unique combination of smaller molecules called amino acids. A gene is a strand of DNA with the instructions for one protein, and its code gives the precise order of amino acids required. Every set of three bases forms a "codon," which translates to one of the 20 amino acids used to build all proteins. A typical protein has about 300 amino acids, and the exact order makes the long, chain-like molecule fold into a specific shape crucial to the protein's function. As the chart here indicates, some of the codons show where one gene stops and the next one starts.

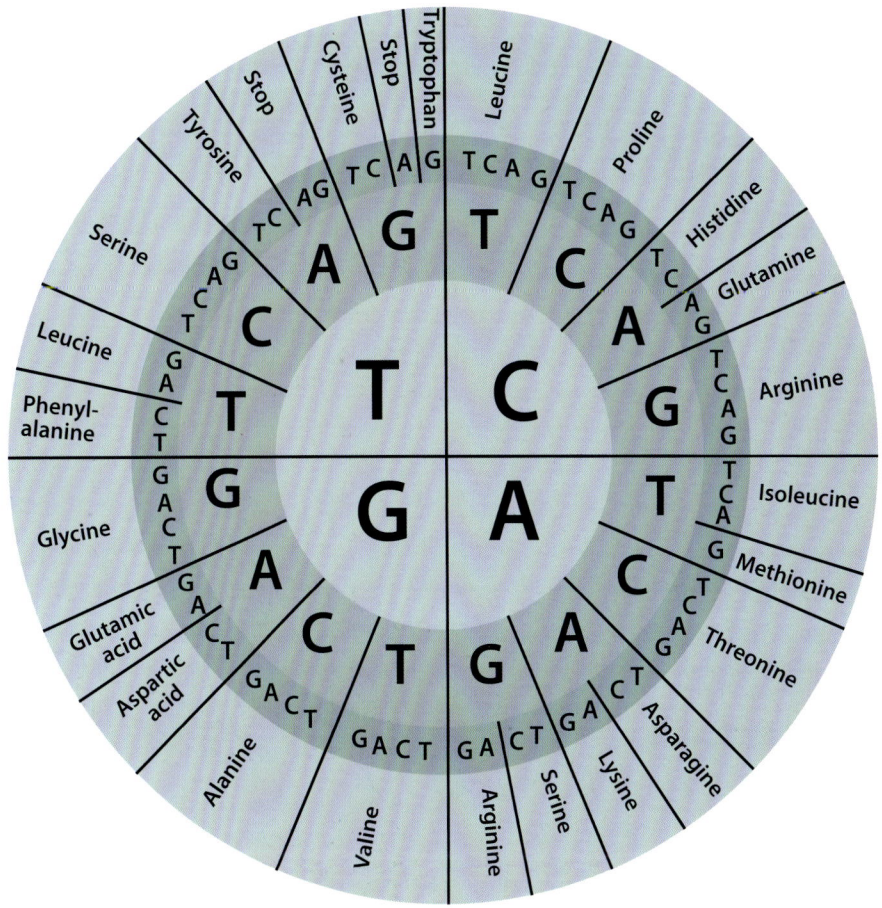

DNA Gene

mRNA

Protein

ncRNA

Function

Function

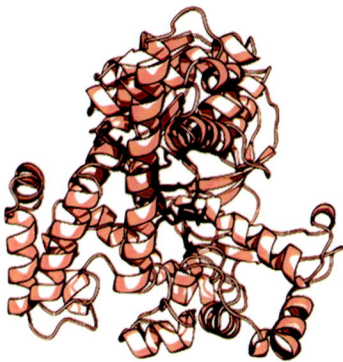

The Central Dogma Biological information always flows from DNA to the protein, never the other way. This is the Central Dogma of Biochemistry. DNA is too fragile to leave a cell's nucleus. Instead, a copy is made on messenger RNA (mRNA). It is that which leaves the nucleus, and is then translated into a protein in a ribosome. These are built mostly from structural or non-coding RNA (ncRNA), which is also copied from another strand of DNA. Within the ribosome, the mRNA's codons match up with those of a transfer RNA (tRNA) molecule. Each tRNA is linked to a specific amino acid, which is then placed into the right part of the protein molecule at the right time.

Computing and electronics

What are semiconductors? Electronics are devices that build on the properties of semiconductors, which are materials that can act as conductors and insulators depending on the conditions. An insulator blocks electric currents because it does not have any free electrons to carry the flow of charge. The electrons are locked in place by the bonds between molecules (a low-energy state called the valence band) and prevented from moving. However, a conductor, such as a metal, has no barrier stopping electrons being free to move and conduct electricity. A semiconductor is half-way between the two: the electrons are locked into the valence band, but, given a small energy boost, will free up to conduct electricity.

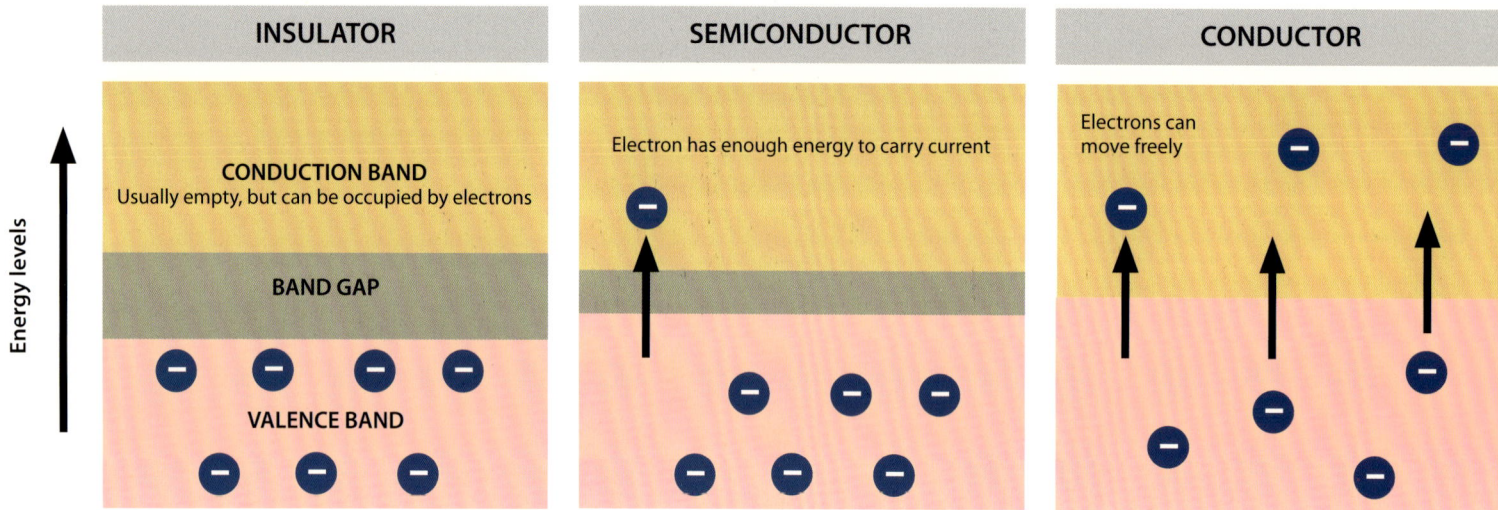

INSULATOR	SEMICONDUCTOR	CONDUCTOR

Energy levels

CONDUCTION BAND
Usually empty, but can be occupied by electrons

BAND GAP

VALENCE BAND

Electron has enough energy to carry current

Electrons can move freely

Computer processing The basic design of a computer has not changed since the late 1940s. The human user controls an input device. That might be a punched card reader, or more likely a keyboard or touchscreen. These inputs are passed to the CPU, where the programming processes them into outputs, such as words being typed on screen. (That's happening before my eyes!) Outputs may be stored, as well, on internal and external drives or uploaded to a data center (or the Cloud).

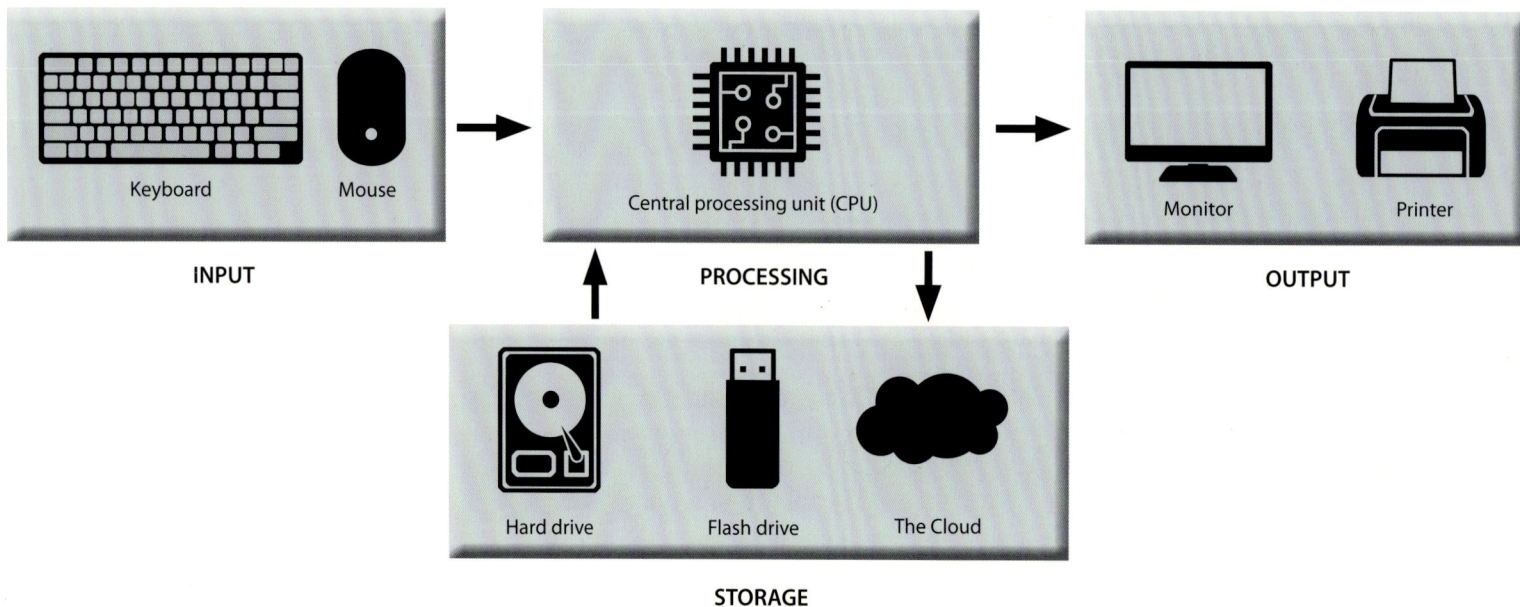

Keyboard   Mouse

INPUT

Central processing unit (CPU)

PROCESSING

Monitor   Printer

OUTPUT

Hard drive   Flash drive   The Cloud

STORAGE

How transistors work Transistors use a sandwich of semiconductors. The "bread" is n-type material which has been "doped" so it contains more negative electrons than normal. The filling is a p-type semiconductor which has few electrons, so more spaces, or "holes," where electrons might fit. At the interface, electrons leak from the n-type to the p-type, creating a "depletion zone." In this state there is no way through the transistor for a current to flow, and so it is switched off. However, when a charge is added to the p-type layer, the electrons are pulled toward it. This creates a bridge of charge between the n-type layers and a current flows through the transistor, so it is switched on.

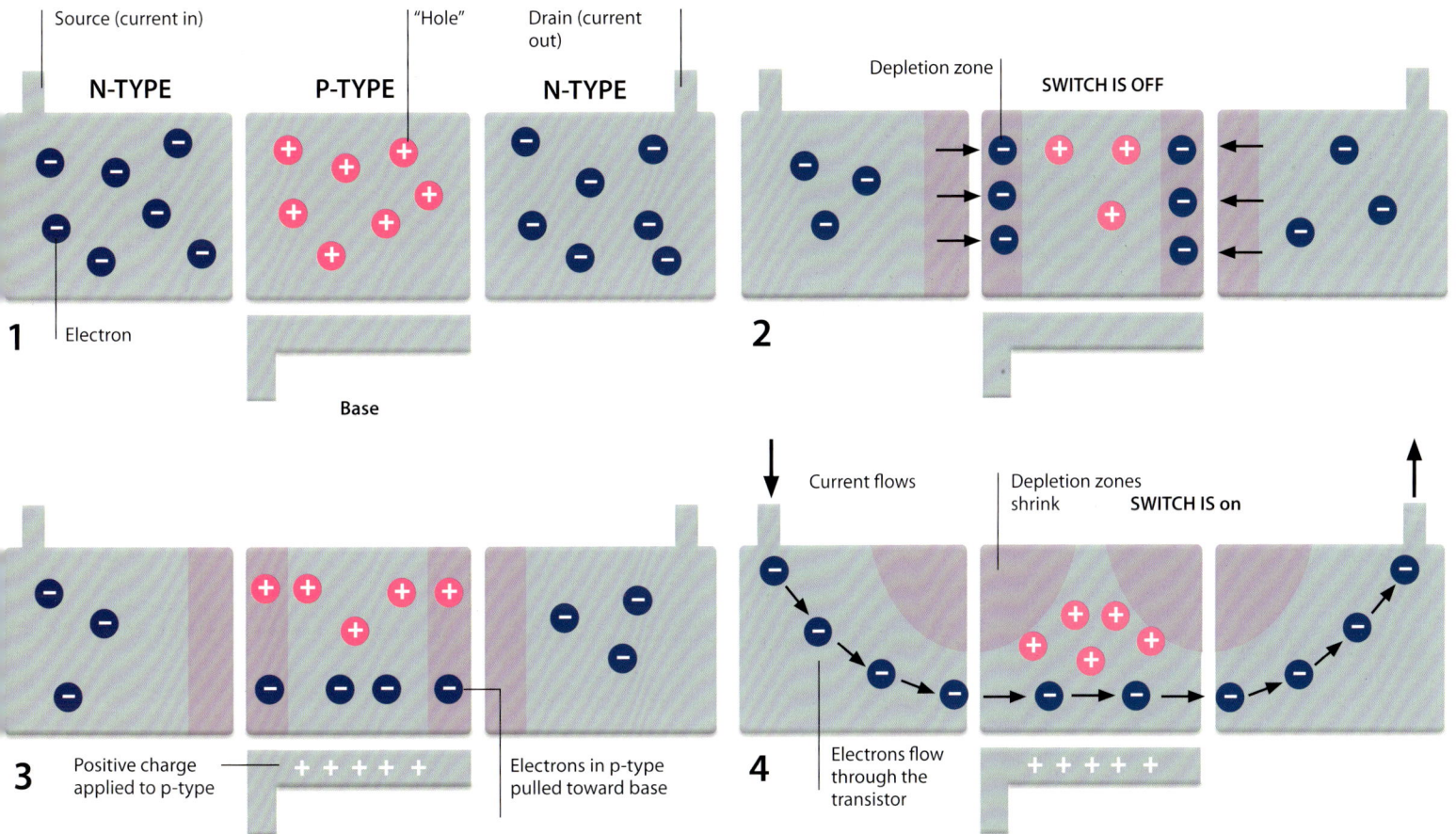

Source (current in)   "Hole"   Drain (current out)

N-TYPE   **P-TYPE**   **N-TYPE**

1 Electron

Base

Depletion zone   **SWITCH IS OFF**

2

3 Positive charge applied to p-type   + + + + +   Electrons in p-type pulled toward base

Current flows   Depletion zones shrink   **SWITCH IS on**

4 Electrons flow through the transistor   + + + + +

Logic gates In 1854, George Boole, an English mathematician, created a form of algebra that could be used to process logical problems. Boolean algebra only works with binary numbers, and the answers it gives are also 0 or 1. In the 1930s, the American Claude Shannon created a system of switches (later made from transistors) that could perform certain kinds of Boolean calculations. These circuits were named logic gates, and they are the means by which data is processed on a microchip today. The charts, right, show five of the simplest gates (those with just one or two inputs) with their truth tables. These show which inputs produce a 1 (for true), and 0 (for false).

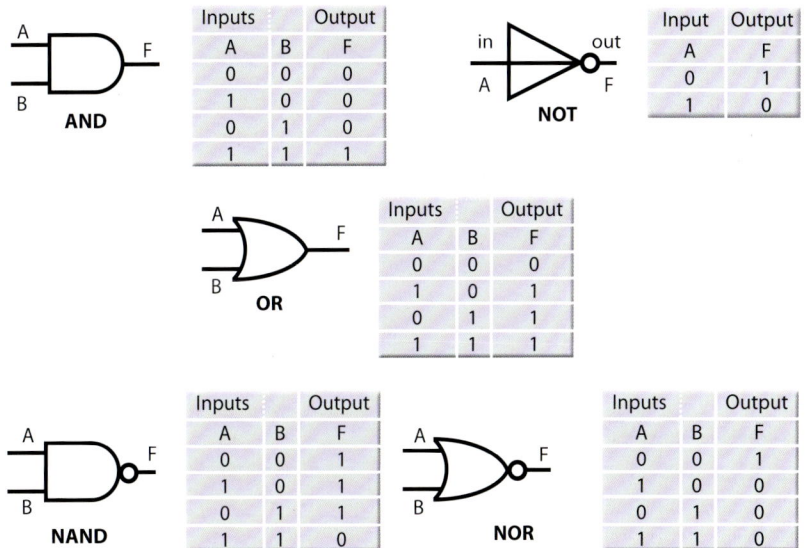

A
B
AND F

Inputs		Output
A	B	F
0	0	0
1	0	0
0	1	0
1	1	1

in → out
A   F
NOT

Input	Output
A	F
0	1
1	0

A
B
OR F

Inputs		Output
A	B	F
0	0	0
1	0	1
0	1	1
1	1	1

A
B
NAND F

Inputs		Output
A	B	F
0	0	1
1	0	1
0	1	1
1	1	0

A
B
NOR F

Inputs		Output
A	B	F
0	0	1
1	0	0
0	1	0
1	1	0

IMPONDERABLES

THE DEVELOPMENT OF TECHNOLOGY WILL NEVER END. IT IS ONE OF THE DEFINING FEATURES OF WHAT WE DO. So there is no doubt that our future selves will be equipped with technology that we can only marvel at, but it is still worth trying to predict what those innovations will be.

What will quantum computing do?

Computing uses billions of on-off switches, and will soon hit a physical size limit. Microchips will have to get bigger—much bigger—to power future tech such as AI. Quantum computers could fix the size problem by using the spooky properties of subatomic particles to process information, and will handle vastly more data at a stroke. We won't run our fridges, or even phones, with quantum chips, but their extraordinary abilities could unlock a way to process information in the vast amounts needed for general artificial intelligence and expert systems.

Quantum computers are still theoretical, but one of the leading contenders is the superconducting D-Wave chip.

How will nanobots work?

Nanotechnology is to do with devices that are less than 100 nanometers, or 100 billionths of a meter, across. Immediately, that brings to mind tiny machines perhaps made from metal and plastic that ape the designs of human-scale engineering. However, down at the nanoscale, engineers build with molecules and individual atoms. Nanotech is already being applied in the form of tiny specks used to coat toughened glass or added to materials at sunny locations to absorb UV light. Engineers are planning more, based on the abilities of complex chemicals such as DNA and proteins which can flex and articulate given the right conditions. So nanotech could be used as levers, screws, cutters, and wheels, plus cages for holding cargos and even gangly-legged bots that crawl. Electronics assembled atom-by-atom could also revolutionize screen and computing technology.

Toothed gears made from tubes of carbon atoms show that simple machines can be put to work even on the smallest scales.

Can we escape the uncanny valley?

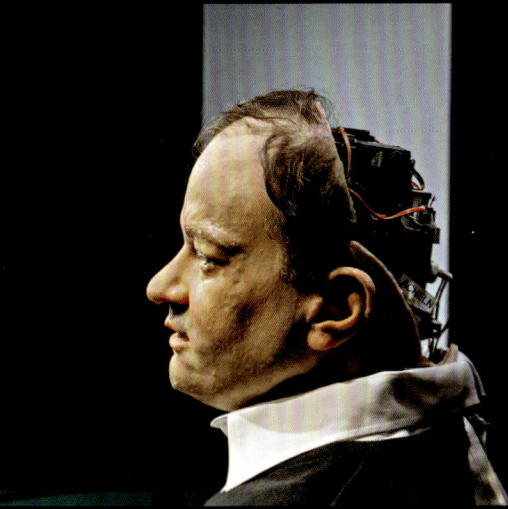

Try as we might, hyper-realistic androids just look eerie, even disgusting. This is the uncanny valley theory. It is probably due to our instinctive fear of death—and the remains of a recently dead human is something to at least be wary of, if not completely terrified. Designers of robots must either create a vital twinkle that makes androids look alive, or we recalibrate our culture to not mind our ghoulish mechanical friends.

Missing the back of your head is something to take notice of if you are human, but an android is not bothered.

Will fusion power ever work?

The promise of nuclear fusion, the power source that supplies the Sun, is a tantalizing one, in that it would provide effectively unlimited and clean energy. The technology needed to control the process—where nuclei smash together and merge—works, but needs to be proven at a large scale, and there are no guarantees. A giant fusion reactor, called ITER, is being built in France, so we will find out in or around 2025.

Fusion reactors are donuts in which beams of nuclei are smashed together.

Is maglev a future technology?

Japanese maglev tests hold the record for fastest land vehicles at 603 km/h (375 mph).

Maglev is short for "magnetic levitation" and that term is most associated with futuristic trains, where, instead of running on old-fashioned wheels, trains float in midair. The levitation is created by superconducting magnets in the train and trackway, and a wave of alternating magnetic polarity pushes the train along at great speed unhindered by friction. It sounds marvelous and thoroughly futuristic. However, despite years of research, the cost of such systems is enormous because the magnets must be kept very cold. Unless there is a breakthrough in superconducting materials, maglev will become a thing of the past.

IMPONDERABLES

Will teleportation machines ever work?

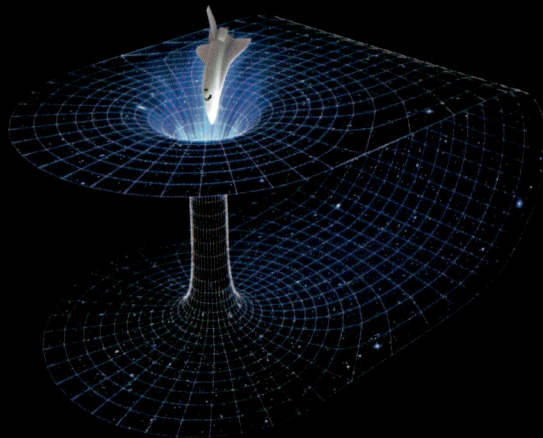

A technology that allows us to travel without moving is not just a plot device in sci-fi stories, it might be possible one day—although perhaps not one day soon. To move faster than light from A to B would need a wormhole connection through hyperspace. That has been predicted by scientists and explored in theory, but few imagine it could be an engineered travel option. (But who knows?) An alternative might be teleportation at the speed of light. This involves transmitting the precise quantum information of an object, such as a traveler's body, to a distant receiver, where it can be assembled as an exact replica. Extracting the quantum information would destroy the original object, so there would be no duplication—the replica becomes the real thing.

Vast amounts of energy would be needed to create warps in space to link distant points as a shortcut.

What is hyperloop?

Elon Musk is well known for his work on rockets and electric vehicles, but he is also exploring a new kind of transportation, which he calls the hyperloop. Envisaged to rival air travel for speed and convenience, it involves propelling passenger cars through long tubes or tunnels. The source of motion is a pressure difference between front and back. Ahead of the car is a near vacuum, and, behind, a pressure wave of air pushes the car along. With low drag forces, the hyperloop is tipped to hit speeds of 1,000 km/h (621 mph), rivaling that of commercial airliners. However, unlike airports, a hyperloop network would connect to the heart of a city. Musk's vision is not entirely new: similar pneumatic railways have been tried since the 1830s. Will he succeed this time?

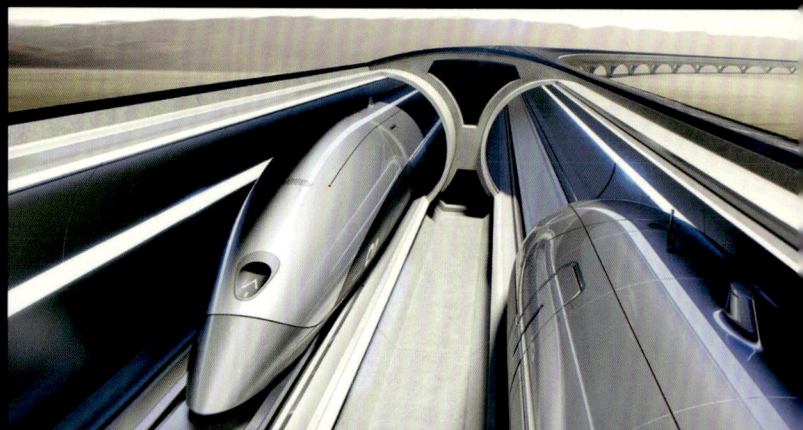

Hyperloop systems are being developed by several rivals.

How do we build space colonies?

The first Mars colony would require its residents to make a one-way journey from Earth.

The drive to establish human colonies off-world began in the 1970s when the first space stations were placed in orbit. That technology—managing air, water, and power onboard—has now been refined and could form the backbone of similar habitats on the Moon or Mars. The difference would be distance. The International Space Station has been permanently crewed since 2000, but it still receives regular shipments of food and water. Supply runs of this kind would be prohibitively expensive for more far-flung outposts. A space colony could not rely solely on solar power like an orbiter, and ideally would have a supply of water that could be split into hydrogen (a fuel) and oxygen (for breathing), and could then be used for washing, drinking, etc. Food would have to be grown in sealed hothouses—once nutrients have been added and toxins removed. Other problems are the lack of air and extreme cold outside the habitats, plus dangerous radiation from sunlight and lung-busting dust particles that could clog machines. No one said it was going to be easy.

What is genetic medicine?

All drugs have a therapeutic window: too much and too little does more harm than good. Modern medicines are deliberately designed to target a particular metabolic pathway, and the therapeutic window seems to vary from person to person. This is due to genetic differences that alter metabolic activity in people, and so the next frontier of pharmacology is to create drugs that are suited to a patient's particular genetic makeup. This might be simply to select the right drug, or it might be to produce personalized drugs that are edited to target a person's metabolism precisely. Genetic medicine promises to make drugs highly effective, but added costs might push the best drugs beyond the reach of many people, unless genetic editing can be automated more

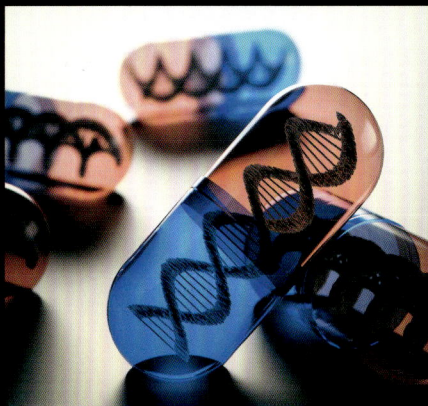

Medical drugs could be targeted to genetic makeup.

The cerebral cortex beneath the skull is electrically active. Is this enough to read minds?

Can we connect our brains together?

Forget keyboards or touchscreens; imagine controlling a device by thinking—by imagining. Already, crude control systems where users learn to control their brain's electrical activity have been trialed. However, the breakthrough will come if we can decode brain activity, and read thoughts as we might outputs from a computer. This is no small task, and involves mapping the brain's many pathways, and applying Big Data analysis to find patterns in the transmissions moving along the paths. If that is possible, we could not only read thoughts but also write them, and record and share directly with another device, including a brain

Is interstellar travel possible?

The fastest object ever made by humans is the Parker Solar Probe. It has hit speeds of 192 km (119 miles) per second, enough to circle Earth in 3.5 minutes. However, even this spacecraft would take 6,300 years to fly to the nearest star. To arrive this year, interstellar explorers would have to have set off 2,000 years before the Great Pyramid was built in ancient Egypt. To have any hope of reaching neighboring stars, we'd need craft that travel at a significant fraction of the speed of light. A solar sail is a flying mirror that picks up speed by being hit by

Interstellar travel might involve entire civilizations spending many centuries en route.

sunlight. The acceleration is gradual, but relentless. A square kilometer (third of a square mile) of sail would reach a twentieth of the speed of light but could only carry a 10-kg (20-lb) cargo. We're going to need a bigger sail.

What would happen if we reach the Technological Singularity?

The Technological Singularity is a hypothetical moment in time when technological progress accelerates beyond the ability of humans to fully comprehend. This is generally expected to happen due to the development of artificial intelligence (AI) that is more intelligent than us. Opinions differ on what would happen next. Some believe that the

This is a problem for the finest minds.

Singularity would be a positive event, leading to a new era of peace and abundance, presumably because the technology is more enlightened than us and so rules in a benign way. The AI is so intelligent that it is able to do everything that ensures society functions. We humans would live a life of leisure. Appealing as it sounds, we would effectively become pets of our AI overlord. Is that so bad? Of course the end of work could lead to widespread unemployment and destitution. It is then an easy step for the quick-thinking AI to regard us not as pets but pests that must be extinguished. Could such a situation ever arise? Who knows, but the smart thinkers, at least the human ones, propose we should examine the issue so we can stop it if needed.

Can materials mend themselves?

Material science links the chemical structure of a substance to its physical properties. It shows that, over time, even the toughest materials accrue tiny cracks and imperfections due to the actions of heat, light, electricity, chemicals, or even sound. Eventually, all artificial materials will fail. Compare this to skin, which is strong, flexible, waterproof, and, if damaged, can self-heal.

Most artificial materials are plastic polymers, which are long chains of repeating molecules (monomers). A self-healing polymer might have an especially reactive monomer, which would be stable when linked with others, but, if the chain were

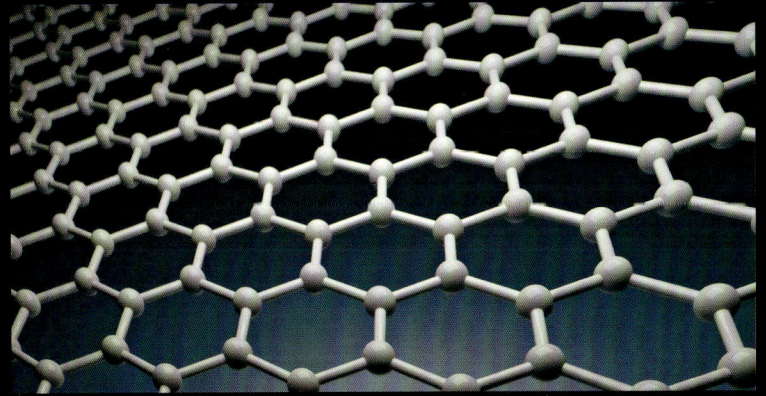

Graphene is a super-strong, manufactured material that recombines when damaged.

broken, it would be able to spontaneously reform the link. Other self-healing materials might release chemicals that fill cracks like a scab heals skin. Nothing artificial will last forever, but self-healing materials will take longer to fail.

Is it possible to make artificial life?

The definitions of life are hotly debated. Perhaps one way to find out is to create a living thing from scratch. That would show if living things are simply the sum of their parts.

Let there be life, but not as we know it now.

Already synthetic life forms—bacteria mainly—have been assembled from the raw materials found in evolved life. These organisms have an artificial genome, edited by their creator, so they represent an entirely new form of life. However, the next step is to replace the evolved materials, such as bi-lipid membranes, DNA and RNA, with engineered structural elements. These elements could be enhanced to be tougher than biological materials or to interact with other technology. Such organisms would be "xenobiological" in that they have no link to a natural ancestor.

Can we use heat to make electricity?

Photovoltaic devices, which transform light energy directly into a current of electricity, are tipped to be a crucial technology as we transition to a low-carbon economy to tackle climate change. However, most of the energy that arrives from the Sun is not light, but invisible infrared—some of which we feel as heat. Infrared radiation carries less energy than light, but, nevertheless, can be transformed into electricity by a thermophotovoltaic effect, potentially giving a much-needed boost to the energy extracted from the Sun's rays. In addition, thermophotovoltaic devices can harvest energy from non-solar sources, such as body heat (using special fabrics in clothing) or heat radiated from buildings, and will even make use of the infrared glow of the night sky.

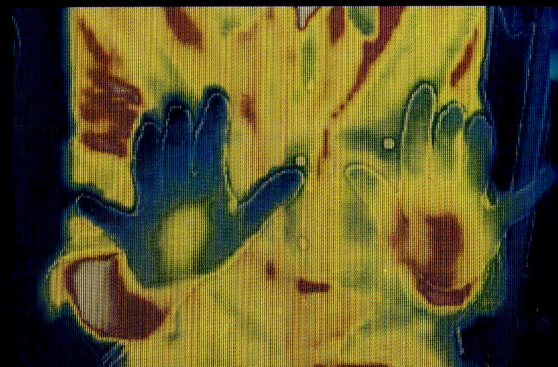

Body heat could be used to power devices.

Is a balloon the easiest way to space?

A helium balloon cannot reach the altitude of an orbiting satellite or space station, but it can lift payloads to heights of 50 km (31 miles). Raising people and cargo to this "near space" zone altitude is much cheaper than by rocket or even an air-breathing aircraft, and so one possible vision of future routes to space is to launch from aerial balloon spaceports. If located at the right height, where winds are low, a gas balloon aircraft can hold a fixed position above the surface, and could be a jumping-off point for catching a rocket to orbit.

The height and lifting power of a balloon is dependent on the volume of gas in the envelope.

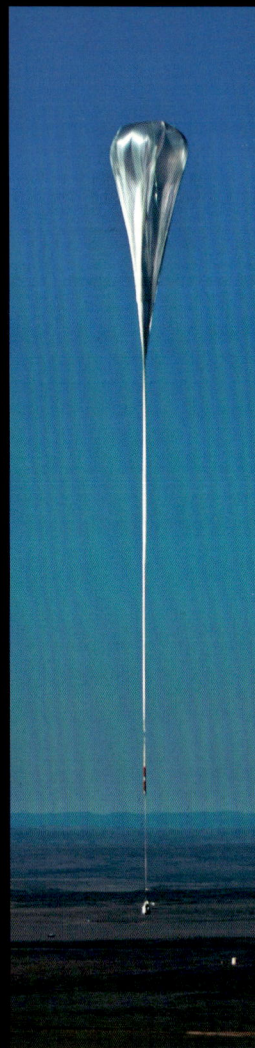

What is the future of the toilet?

Today we flush waste with water clean enough to drink—while, of course, hundreds of millions of us have no access to a flush toilet. These two extremes may result in a redesigned toilet which does not use water, but instead converts solid waste into safe and useful fertilizer or fuel.

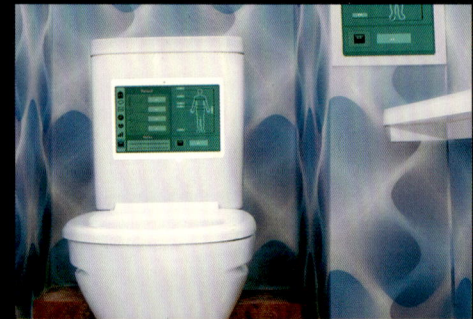

Biosensors in the bowl will also be able to screen waste for signs of illness—and send details to doctors.

Will future farms float on the ocean?

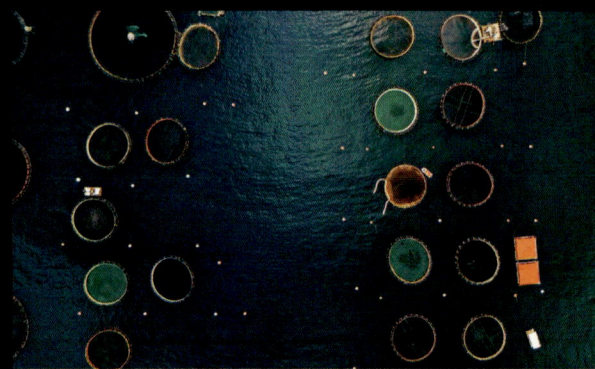

Roughly a third of our planet's dry land is covered by farmland. Another third is forest, and the rest is too barren for either. Climate change means that we need to increase forest cover, and so if we are to grow the food we need, then the idea of "farmland" will need to broaden. Already, fish, such as salmon, are farmed in open seawater enclosed by netting, so why not extend that idea to terrestrial livestock? In 2019, the first floating cowshed set sail from Rotterdam with 34 dairy cows aboard. Perhaps floating hothouses, vineyards, and orchards will next be moored in tropical waters.

The open ocean might become the farmland of the future.

Can we breed back extinct species?

Most of the life-forms that have evolved on Earth are now extinct. However, some extinctions are not quite as final as others. After all, the dinosaurs never went extinct: in fact, a lot of us eat one every Thanksgiving. In other words, birds are the living descendants of dinosaurs. This is called pseudoextinction and differs from true extinction, which is when creatures, such as trilobites, are wiped out and leave no descendants. So, bird DNA shares a lot with dinosaur DNA, most especially with the therapods like velociraptors and *T. rex*. Is it possible to alter this DNA so, instead of a bird, it builds a dinosaur, rescuing it from extinction? There are two possible approaches. The first is artificial breeding, already underway to bring back the quagga, an extinct subspecies of zebra, from its living relatives. But, this technique would take thousands of generations to breed a dinosaur from a turkey. The alternative is to redesign the turkey DNA to produce something like a dinosaur, which requires a more in-depth understanding of bird—and dinosaur—genomes. And would the finished animal really be the extinct species or just something that looks like it?

Woolly mammoths are close relatives of Asian elephants. Could we breed them back?

Will carbon-capture technology make a difference?

Human activity, such as burning fossil fuels, is increasing the amount of carbon dioxide in the air. For decades it has been clear that this change boosts the amount of the Sun's energy trapped in the atmosphere—and very few climate scientists dissent from the view that this energy will create rapid and extreme shifts in our world's weather systems. The obvious thing to do is reduce the amount of carbon dioxide in the air. The simplest mitigation is to plant 1.2 trillion trees (that's 150 each) and lock away the atmospheric carbon as wood. But even that is not enough; at least in the medium term we will need an artificial technology that captures carbon dioxide from the air. There are several systems being tested that use chemicals and filters, and whatever works (if it works) would need to be rolled out on a vast scale for decades. The next question is what to do with all the captured carbon. A hybrid approach that uses technology and natural processes is to pump iron fertilizers into the oceans. That would stimulate the growth of plankton, which would then die and sink to the seabed, building a carbon sink of sediment. The other big idea is to store sequestered carbon in empty oil fields.

Oil and gas fields are good places to store unwanted greenhouse gases.

Great Inventors

A LIST OF GREAT INVENTORS IS A LONG ONE, AND NEARLY IMPOSSIBLE TO REDUCE TO THE TOP **31,** but here goes! Inventors have a mix of genius, curiosity, and tenacity, which helps them see problems that need solving, and gives them the means to make their ideas a reality. Inventors are not just born, but made, in that many of today's tech innovators grew up surrounded by computers—and went on to create a world where we are all surrounded by them, as well! The mark of a great inventor is that their work remains relevant into the future. You may not have heard of these people but you'll know what they created.

Vitruvius

Born	1st century BCE
Place of birth	Fondi, Italy
Died	1st century BCE
Importance	Influential architect

From the Renaissance to the 18th century, the Roman architect and engineer Vitruvius was the most important authority on neoclassical architecture throughout Europe. Having served as a military engineer, Vitruvius became skilled at building in the classical Greek style, which the Romans inherited. Vitruvius is most celebrated for his 10-volume treatise *De architectura* (On Architecture), which laid out practical information about building materials, city planning, and home decoration. However, the work also ranged across a wide variety of subjects, from clocks to astronomy and siege engines.

Archimedes

Born	Born c.287 BCE
Place of birth	Syracuse, Sicily
Died	c.212 BCE
Importance	Inventor and mathematician

Renowned as a great mathematician, Archimedes was also a talented inventor who devised both the compound pulley and the Archimedes screw, a hydraulic device for lifting water that is still in use today. During the Roman conquest of Sicily in 214 BCE, Archimedes designed weapons, including a huge claw that

lifted enemy ships out of the water, and a glass or mirror that set them on fire by focusing the Sun's rays. His lasting claim to fame is his cry of "Eureka!" when he stepped into his bath and realized that the displacement of water could be used to measure volume and density.

Heron of Alexandria

Born	c.10 CE
Place of birth	Alexandria, Egypt
Died	c.70 CE
Importance	Prolific Roman inventor

Heron, or Hero, lived in Egypt under the rule of the Romans. Today, he is known for the aeolipile, an early form of steam turbine probably designed as a toy or novelty rather than as a serious attempt to generate power. He also invented practical devices, including a water organ, an organ powered by wind, a fire engine, automatic doors, and a coin-operated machine to sell holy water at temples. He wrote books on pneumatics, mechanics, measurements, and optics, and made a number of breakthroughs in mathematics, including Heron's formula for calculating the area of a triangle from the length of its sides.

Zhang Heng

Born	78 CE
Place of birth	Nanyang, China
Died	139 CE
Importance	Inventor of the seismograph

Zhang Heng was an astronomer, mathematician, engineer, painter, and poet at the court of the Eastern Han dynasty during a period when China reached a peak of cultural and political achievement. Today, he is remembered chiefly as the inventor of the seismograph, but during his lifetime Zhang was more celebrated as a poet. Around 112 CE he was appointed an official at the Emperor's court, eventually becoming chief astrologer and minister, in which role he undertook the reform of the calendar—a crucial task, because the Chinese believed that civil life was reflected and controlled by the heavens.

John Harrison

Born	April 3, 1693
Place of birth	Foulby, Yorkshire, England
Died	March 24, 1776
Importance	Inventor of chronometers for marine navigation

The self-taught Yorkshire clockmaker John Harrison solved one of the scientific challenges of his age: inventing a timepiece so accurate that it would allow the accurate calculation of longitude at sea. Having been building clocks since the age of 20, Harrison traveled to London to gain initial support for building a "sea clock." It took decades before Harrison's second timepiece passed the stringent tests laid down by the British Admiralty. Under pressure from King George III, Parliament finally agreed to pay Harrison a handsome sum that made him a wealthy man for the final years of his life.

Otto von Guericke

Born	November 30, 1602
Place of birth	Magdeburg, Germany
Died	May 11, 1686
Importance	Inventor of the vacuum pump

A leading German scientist of the Enlightenment, Otto von Guericke studied math and physics at universities in Germany and Holland. His father and grandfather had both served as mayors of his home town, Magdeburg, but von Guericke's own career there was disrupted by the Thirty Years' War. He restored his fortunes through brewing and, in time, became mayor himself, in which role he attempted to gain Magdeburg recognition as a free city. In addition to his well-known work on air pumps and vacuums, von Guericke was also known for his investigations into static electricity and the nature of space, which he proposed was a vacuum.

William Cullen

Born	April 15, 1710
Place of birth	Hamilton, Lanarkshire, Scotland
Died	February 5, 1790
Importance	Inventor of refrigeration

William Cullen belonged to a generation of influential Scottish thinkers in what is known as the Scottish Enlightenment. Trained as an apothecary surgeon, he later became a highly influential chemistry professor in Glasgow and Edinburgh. Cullen's students went on to make significant contributions to science. Cullen did not pursue his discovery of mechanical refrigeration himself, and so he is more widely recognized as an educator than as an innovator. He was particularly interested in the effects of the nervous system, and developed a theory called nosology that was an attempt to classify diseases.

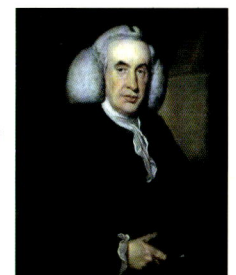

Alessandro Volta

Born	February 18, 1745
Place of birth	Como, Italy
Died	March 5, 1827
Importance	Inventor of the battery

Volta came from a noble family in Como in northern Italy, where he became a professor of physics and soon made his name by promoting a device that produced static electricity (although it was not actually his invention). When Luigi Galvani announced the discovery of "animal electricity," which he claimed was made in living things, Volta instead used Galvani's findings to invent the first battery, the voltaic pile. This was based on the exchange of current between two different metals in a liquid electrolyte. Volta's achievements were widely recognized during his lifetime, and he was made a count by Napoleon Bonaparte.

Nicolas Appert

Born	November 17, 1749
Place of birth	Châlons-en-Champagne, France
Died	June 1, 1841
Importance	Inventor of food canning

Like many other inventors, the Parisian sweet-maker Nicolas Appert was encouraged to turn inventor by the offer of a financial reward—12,000 francs to find a way to preserve food for armies campaigning under the French general, and later emperor, Napoleon Bonaparte. Unlike most other inventors, however, Appert had little understanding of how the method he developed worked, even though he successfully claimed the reward. After being granted the money in 1809, Appert wrote a book about his preservation process and used his money to open the world's first canning factory. Early canning was often known as "appertization."

Edward Jenner

Born	May 17, 1749
Place of birth	Berkeley, Gloucestershire, England
Died	January 26, 1823
Importance	Pioneer of vaccination

Jenner studied with a local doctor near his home in Gloucestershire before training as a surgeon in London, then returning home to serve as a family doctor. Jenner was a keen natural historian, and his understanding of animal biology encouraged his interest in vaccination. Jenner later gave up his medical duties to further develop safe vaccination, allowing the government to ban the riskier practices such as variolation (rubbing smallpox pus into scratches). Jenner was widely celebrated for his achievement and was rewarded with a government pension and various academic positions.

Nicéphore Niépce

Born	March 7, 1765
Place of birth	Chalon-sur-Saône, France
Died	July 5, 1833
Importance	Took the earliest surviving photograph

Joseph Niépce, who adopted the name Nicéphore while studying religion, was an officer in the armies of Napoleon Bonaparte. He later became administrator of the district of Nice, reportedly resigning because he was unpopular. With his brother, Claude, Niépce developed the first internal combustion engine, named the Pyréolophore. In addition to the pioneering experiments with photography for which he is remembered today, Niépce and his brother also invented a hydraulic water pump, and Nicéphore built an early type of bicycle, the velocipède. Both Claude and Nicéphore died in poverty as a result of misguided business ventures based on the Pyréolophore.

Michael Faraday

Born	September 22, 1791
Place of birth	Newington Butts, Surrey, England
Died	August 25, 1867
Importance	Prolific scientific discoverer and popularizer

Coming from a poor background, Michael Faraday had little schooling, but taught himself by reading during his apprenticeship as a bookbinder in London. He was interested in science from an early age, and the notes he took at lectures by Humphrey Davy led to him being offered a post as Davy's chemistry assistant at the Royal Institution. Although his career suffered because of his lack of social status, Faraday became highly influential for his work on the relationship between electricity and magnetism. He began the Christmas Lectures at the Royal Institution in London, a tradition that continues today to present science to children.

Ada Lovelace

Born	December 10, 1815
Place of birth	Piccadilly, Middlesex, England
Died	November 27, 1852
Importance	Inventor of the first computer program

Daughter of the notorious Romantic poet Lord Byron, Ada Lovelace was taught math from an early age, partly because her mother wanted Ada to avoid the madness believed to have afflicted Byron. Having married the Earl of Lovelace, Ada met Charles Babbage, inventor of the Analytical Engine for performing calculations.
Invited by Babbage to translate an Italian article about one of his lectures, Lovelace added extensive ideas of her own, including a way of programming the engine to compute Bernoulli numbers. Although it was never built, she is widely celebrated as the inventor of the first computer program.

Joseph Henry

Born	December 17, 1797
Place of birth	Albany, New York, USA
Died	May 13, 1878
Importance	Pioneer of electrical devices

Joseph Henry discovered the principle of self-induction, but his improvements to the electromagnet were overshadowed by Michael Faraday's discovery of electromagnetic induction in 1831. As a professor at what became Princeton University, Henry was a pioneer of the transformer, experimented with radio waves, and helped develop the electric telegraph, promoting the system of Samuel Morse. Having advised Union forces during the American Civil War and laid the foundations of the U.S. Weather Service, Henry also helped organize the National Academy of Sciences. The international unit of inductive resistance, the henry, is named in his honor.

Alfred Nobel

Born	October 21, 1833
Place of birth	Stockholm, Sweden
Died	December 10, 1896
Importance	Inventor of dynamite

It is said that Alfred Nobel, the inventor of dynamite, founded the Nobel Prizes for scholarship after a French newspaper prematurely published his obituary. The article began, "Dr. Alfred Nobel, who became rich by finding ways to kill more people faster than ever before, died yesterday." Nobel excelled at chemistry and founded a business manufacturing explosives and armaments. An explosion that killed his younger brother drove Nobel to invent safer alternatives, dynamite and later gelignite. These made Nobel highly successful in a period when warfare, mining, and railroad construction created a high demand for explosives.

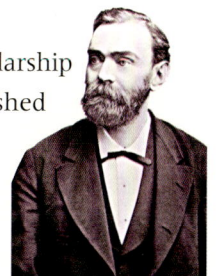

Karl Benz

Born	November 25, 1844
Place of birth	Karlsruhe, Baden, Germany
Died	April 4, 1929
Importance	Inventor of the first automobile

Karl Benz was brought up in poverty after his father, a locomotive driver, died in an accident. He studied mechanical engineering at the University of Karlsruhe. With the financial support of his wife, Bertha, Benz improved the internal combustion engine and then developed the world's first functioning car. The Benz Motorwagen of 1885 incorporated many Benz inventions including the car battery, clutch, and spark plugs. It was popularized by a trip Bertha made to visit her mother. In 1926, Benz's company merged with rivals Daimler to produce cars under the brand Mercedes-Benz, with the name Mercedes in honor of the daughter of a Daimler executive.

Alexander Graham Bell

Born	March 3, 1847
Place of birth	Edinburgh, Scotland
Died	August 2, 1922
Importance	Inventor of the telephone

The son of a deaf mother and a father who taught deaf children, Alexander Graham Bell was fascinated from an early age by sound and hearing. Bell left school at age 15, and moved to Canada. Having followed his father into teaching, Bell later changed career to concentrate on improving the telegraph, and developed the telephone with his assistant, Thomas A. Watson. The patent Bell was granted for the telephone, even before he had a functioning device, made him a wealthy man. He fostered the spread of scientific knowledge—he was a founder of the National Geographic Society and a supporter of the journal *Science*.

Thomas Edison

Born	February 11, 1847
Place of birth	Milan, Ohio, USA
Died	October 18, 1931
Importance	Prolific developer of electrical devices

Edison trained as a telegraph operator on the American railway system before the rapidly changing technology of telegraphy inspired him to go on to become one of the most prolific inventors in history, with more than 1,000 patents to his name. Having given up school early because of his partial deafness, he later stressed the importance of persistence over genius for an inventor: "Any other bright-minded fellow can accomplish as much if he will stick like hell and remember nothing that's any good works by itself." In 1876, Edison built a laboratory at Menlo Park, New Jersey, where the phonograph, cinema camera, and incandescent light bulb were invented.

Nikola Tesla

Born	July 10, 1856
Place of birth	Smiljan, now Croatia
Died	January 7, 1943
Importance	Prolific electrical engineer

The son of a Serbian Orthodox priest, Nikola Tesla studied electrical engineering at the Polytechnic Institute at Graz, Austria, where he became interested in the potential of alternating current (AC) electricity. Having fallen out with his professors, Tesla dropped out and sailed for New York City, where he arrived in 1884 with just 4 cents in his pocket. Tesla worked with Thomas Edison until the two fell out, and Tesla then sold many of his ideas about AC generation to Edison's rival, George Westinghouse. Tesla went on to develop more electrical and radio devices, but his terrible business sense left him bankrupt when he died in a New York hotel.

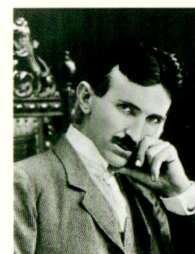

Grace Hopper

Born	December 9, 1906
Place of birth	New York City, USA
Died	January 1, 1992
Importance	Developed early programming language

After a spell as a mathematics professor at Vassar College in New York, Hopper tried to join the U.S. Navy during World War II, but was only let into the Reserves. In 1944, she was involved in the development of the Harvard Mark 1 computer, and this led to further work on the UNIVAC, the first commercial computer. Hopper built the compiler, the software that created the machine code, or 1s and 0s, that the computer understood. Her system converted English words into code, so programs could be written using words. This led Hopper to be a key figure in the 1950s' development of COBOL, one of the first high-level programming languages.

Theodore Maiman

Born	July 11, 1927
Place of birth	Los Angeles, USA
Died	May 5, 2007
Importance	Inventor of the laser

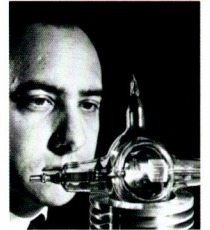

The son of an electrical engineer, Ted Maiman began fixing electronic devices at the age of 12 before studying physics at Stanford University. While working at the Hughes Research Laboratory, he became interested in the possibility of using rubies to amplify light. Having succeeded, Maiman found his paper on the laser rejected by journals but finally got it published in the British journal *Nature*. The article was later described as "probably more important per word than any of the papers published by *Nature* over the last century." Maiman later set up companies to exploit the commercial possibilities of lasers.

William Shockley

Born	February 13, 1910
Place of birth	London, England
Died	August 12, 1989
Importance	Co-inventor of the transistor

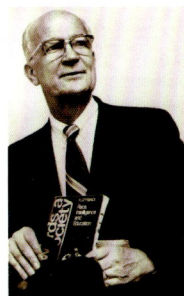

The son of mining engineers, William Shockley studied physics at Caltech and MIT before joining the celebrated Bell Laboratories to carry out research into electronics. He led the team that won the Nobel Prize for developing the transistor. Notoriously difficult to work with, Shockley fell out with his collaborators, and moved to Mountain View, California, to be near his mother. There, he set up the first electronics plant in what became known as Silicon Valley. Many people who once worked for him led the area's involvement in electronics and computing. Shockley became increasingly controversial for his views on race and intelligence.

Geoffrey Hinton

Born	December 6, 1947
Place of birth	London, England
Died	-
Importance	Pioneer of artificial intelligence

Geoffrey Hinton is considered to be the founding figure of deep learning, the form of machine learning that has created modern AIs. The British-Canadian's involvement in this field began through his interest in cognitive psychology, when he attempted to learn about the brain by making a computer that did the same things. His greatest contribution to AI was development of backpropagation, an algorithm that allows neural networks to learn patterns in data. He is one of the founders of the Google Brain project, which he left in 2023 so he could advocate more widely for government and industry regulation of artificial intteligence.

Steve Jobs

Born	February 24, 1955
Place of birth	San Francisco, USA
Died	October 5, 2011
Importance	Founder of Apple computers

In his trademark black turtleneck sweater, Steve Jobs became the stereotype of the technology entrepreneur of the late 20th century. Adopted as a child, Jobs grew up in California, where he became interested in oriental philosophy before dropping out of college to found Apple Inc. in 1976. The Apple computer was based on the one invented by his business partner, Steve Wozniak. Under Jobs' leadership, Apple became a major player in computing and communication technology, with products including the iPhone and the iPad. He also developed computer animation with the movie studio Pixar. Jobs died from cancer aged just fifty-six.

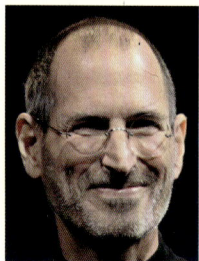

Bill Gates

Born	October 28, 1955
Place of birth	Seattle, Washington, USA
Died	-
Importance	Co-founder of Microsoft

Bill Gates wrote his first computer program at the age of 13 on his school's new computer. After studying at Harvard, he dropped out in order to found a software company called Microsoft with his colleague Paul Allen. They wrote the BASIC program for the Altair 8800, before going on to supply the DOS operating system for the IBM PC and then writing the software for the Windows operating system. Microsoft made Gates one of the world's wealthiest men, and he and his wife, Melinda, set up the world's largest charitable foundation to promote public health in the developing world. Gates also provides personal backing for a range of other philanthropic activities.

Tim Berners-Lee

Born	June 8, 1955
Place of birth	London, England
Died	-
Importance	Inventor of the World Wide Web

With parents who were both pioneers of computing science in the decades after World War II, it was almost inevitable that Tim Berners-Lee would follow in their footsteps. Working at CERN in Geneva, Berners-Lee later had the idea of combining hypertext technology with domain names to create what he called the World Wide Web. In 1990, Berners-Lee built the world's first website, in which he described the Web. It was later selected as the most important cultural moment in world history. Berners-Lee made the Web available free for all users, without royalties. Since his invention of the Web, Berners-Lee has promoted Web standards.

Jennifer Doudna

Born	February 19, 1964
Place of birth	Washington, D.C., USA
Died	–
Importance	Developed CRISPR gene editing system

Moving with her academic parents to Hawaii as a child, Doudna's interest in the natural world was inspired by the tropical wildlife of her new home. She studied biochemistry in college, and began a career in researching the biochemistry of RNA and the way it works inside cells. This work took Doudna to Harvard Medical School and then Berkeley. In 2012, Doudna led a team that created the CRISPR toolkit for editing DNA (although there are rival claims from other institutions). In 2017, Doudna founded Mammoth Biosciences to apply her expertise. During the 2020 COVID-19 pandemic, the company used CRISPR to develop diagnostic tools.

Larry Page

Born	March 26, 1973
Place of birth	East Lansing, Michigan, USA
Died	-
Importance	Co-founder of Google

Famous as one of the co-founders of Google along with Sergey Brin, Larry Page grew up in a home full of computers as both his parents were among the early generation of computer scientists. Page began using computers at age six, in addition to proving a talented musician. After graduating from the University of Michigan, Page enrolled for a PhD at Stanford, where he involved Brin in his search for developing a method of ranking Web pages. As CEO of Google, Page became notorious for his relentless desire to improve the search engine. He announced he was stepping down from his role in 2019.

Satoshi Nakamoto

Born	April 5, 1975
Place of birth	Japan
Died	-
Importance	Creator of Bitcoin

This is the name given to the inventor of the online currency Bitcoin that launched in 2009. No one knows who they are, but, despite this, the inventor has not only established themselves as a computer pioneer, but has also become one of the world's fifty richest people thanks to holdings of the cryptocurrency. The only clues to Nakamoto's identity come in the content and timing of their online posts about Bitcoin, which have led some experts to conclude that they are not Japanese but more likely to be English-speaking and possibly European. A number of potential individuals have been suggested as Nakamoto, but none with overwhelming conviction.

Sergey Brin

Born	August 21, 1973
Place of birth	Moscow, Russia
Died	-
Importance	Co-founder of Google

Moving with his parents from the communist Soviet Union, Sergey Brin arrived in Maryland at the age of six and followed his father in studying mathematics and computer science, first in Maryland and then at Stanford. Early in his postgraduate career at Stanford, Brin met Larry Page, with whom he created an algorithm to rank websites in order of their likely importance. Initially constructed in the pair's college dorm rooms, the service later became Google. Google made Brin one of the world's richest people, and enabled him to pursue his own interests, including using new energy and vehicle technologies to tackle issues such as climate change.

Mark Zuckerberg

Born	May 14, 1984
Place of birth	White Plains, New York, USA
Died	-
Importance	Co-founder of Facebook

At the age of just 12, Mark Zuckerberg built a computer messaging service to allow his dentist father to be informed by reception when his patients arrived. Zuckerberg remained fascinated by computers, building games and music software, and receiving education from a private tutor. At Harvard, he developed software to help students to select their courses and also a social networking project called Harvard Connection. This prompted Zuckerberg to start Thefacebook, initially run from a dorm room before Zuckerberg moved to California in 2004 to run the site full-time, as Facebook.

BIBLIOGRAPHY AND OTHER RESOURCES

Books

Billington, David P. and David P. Billington Jr. *Power, Speed, and Form: Engineers and the Making of the Twentieth Century.* Princeton University Press, 2006

Brown, David J. Bridges: *Three Thousand Years of Defying Nature.* Firefly, 2005

Challoner, Jack. *1001 Inventions That Changed the World.* B.E.S. Publishing, 2009

Foreman, Louis and Jill Gilbert Welytok. *The Independent Inventor's Handbook: The Best Advice from Idea to Payoff.* Workman Publishing Company, 2009

Gibby, Darin. *Why Has America Stopped Inventing?* St. Martin's Press, 2011

Inventions: A Visual Encyclopedia. DK Children, 2018

Johnson, Steven M. *Have Fun Inventing: Learn to Think Up Products and Imagine Future Inventions.* BookSurge Publishing, 2012

Kidder, Tracy. *The Soul of a New Machine.* Back Bay Books, 2005

Maeda, John. *How to Speak Machine: Computational Thinking for the Rest of Us.* Portfolio/Penguin, 2019

Munroe, Randall. *Thing Explainer: Complicated Stuff in Simple Words.* Houghton Mifflin Harcourt, 2015

O'Mara, Margaret. *The Code: Silicon Valley and the Remaking of America.* Penguin, 2020

Russell, Stuart. *Human Compatible: Artificial Intelligence and the Problem of Control.* Viking, 2019

Seabrook, John. *Flash of Genius And Other True Stories of Invention.* St. Martin's Griffin, 2008

Siebel, Thomas M. *Digital Transformation: Survive and Thrive in an Era of Mass Extinction.* RosettaBooks, 2019

Smith, Brad and Carol Ann Browne. *Tools and Weapons: The Promise and the Peril of the Digital Age.* Penguin, 2019

Srinivasan, Ramesh. *Beyond the Valley: How Innovators around the World are Overcoming Inequality and Creating the Technologies of Tomorrow.* MIT Press, 2019

Apps

ARCore Elements

AutoCAD 360

Basic Engineering Dictionary

Discovery & Inventions News

iCircuit

iF design

Kramer KRC Reality Viewer

Master Class AutoCAD Edition

Materials & Processes

Robotics Engineering

Sketchdrive

TurboViewer

Archives

Ada Lovelace Archive and Manuscripts, Bodleian Library, University of Oxford, Oxford, England

Alexander Graham Bell Family Papers, Library of Congress, Washington, DC, USA

Alfred Nobel Family Archive, National Archives, Stockholm, Sweden

Consultation Letters of Dr William Cullen, Royal College of Physicians of Edinburgh, Edinburgh, Scotland

Grace Murray Hopper Collection, Archives Center, National Museum of American History, Smithsonian Institution, Washington, DC, USA

Papers of Joseph Henry, Smithsonian Institution, Washington, DC, USA

Papers of William Cullen, Edinburgh University Library, Edinburgh, Scotland

Papers of William Cullen, University of Glasgow Special Collections, Glasgow, Scotland

Theodore Maiman Papers, Niels Bohr Library & Archives, American Institute of Physics, College Park, Maryland, USA

William Shockley Papers, Stanford University Libraries, California, USA

Museums and Places to Visit

Air Defence Radar Museum, Norfolk, England

Birla Industrial and Technological Museum, Kolkata, India

Cade Museum for Creativity and Invention, Gainesville, Florida, USA

California Academy of Sciences, San Francisco, California, USA

City of Science and Industry, Paris, France

Computer History Museum, Mountain View, California, USA

Cosmonautics and Aviation Center, Moscow, Russia

Deutsches Museum, Munich, Germany

Discovery Museum, Newcastle upon Tyne, England

Dr. Jenner's House, Berkeley, Gloucestershire, England

Exploratorium, San Francisco, California, USA

Faraday Museum, Royal Institution, London, England

German Museum of Technology, Berlin, Germany

Hong Kong Space Museum, Hong Kong, China

Ingenium, Ottawa, Ontario, Canada

Inventionland, Pittsburgh, Pennsylvania, USA

Israel National Museum of Science, Technology, and Space, Haifa, Israel

James Clerk Maxwell House, Edinburgh, Scotland

John Gorrie Museum State Park, Apalachicola, Florida, USA

Lemelson Hall of Invention and Innovation, Smithsonian National Museum of American History, Washington, DC, USA

Maloka Museum, Bogotá, Colombia

Marconi Centre Museum, The Lizard, Cornwall, England

Mirador Interactive Museum, La Granja, Chile

Miraikan (The National Museum of Emerging Science and Innovation), Tokyo, Japan

MIT Museum, Cambridge, Massachusetts, USA

Museo Nazionale Scienza e Tecnologia "Leonardo da Vinci," Milan, Italy

Museum for Old Techniques, Grimbergen, Belgium

Museum of Applied Arts and Sciences, New South Wales, Australia

Museum of Arts and Crafts, Paris, France

Museum of Ideas and Inventions, Barcelona, Spain

Museum of Innovation and Science, Schenectady, New York, USA

Museum of Technology, Helsinki, Finland

Museum of Technology, Warsaw, Poland

Museum of the Future, Dubai, United Arab Emirates

NASA Visitor Centers, various locations, USA

National Air and Space Museum, Smithsonian Institution, Washington DC, USA

National Inventors Hall of Fame and Museum, Alexandria, Virginia, USA

National Museum of Computing, Bletchley Park, Buckinghamshire, England

National Museum of Science and Technology, Madrid, Spain

National Technical Museum, Prague, Czech Republic

NEMO Science Museum, Amsterdam, The Netherlands

Nicéphore-Niépce House Photo Museum, Saint-Loup-de-Varennes, France

Nikola Tesla Technical Museum, Zagreb, Croatia

Norwegian Museum of Science and Technology, Oslo, Norway

Ontario Science Centre, Toronto, Ontario, Canada

Otto-von-Guericke-Museum, Magdeburg, Germany

Palace of Running Water, Buenos Aires, Argentina

Polytechnic Museum, Moscow, Russia

PUCRS Museum of Science and Technology, Porto Alegre, Brazil

Quito Astronomical Observatory, Quito, Ecuador

Royal Observatory, Greenwich, London, England (John Harrison's timekeepers)

Science Museum, London, England

Science Museum of the University of Coimbra, Coimbra, Portugal

Science Space (Espacio Ciencia/LATU), Montevideo, Uruguay

Shanghai Science and Technology Museum, Shanghai, China

Slovak Technical Museum, Košice, Slovakia

SPARK Museum of Electrical Invention, Bellingham, Washington, USA

Swedish National Museum of Science and Technology, Stockholm, Sweden

Swiss Museum of Transport, Lucerne, Switzerland

Technik Museum Sinsheim / Technik Museum Speyer, Germany

Technoseum, Mannheim, Germany

The Tech Interactive, San Jose, California, USA

Thomas Edison National Historical Park, West Orange, New Jersey, USA

Universeum, Gothenburg, Sweden

Universum, Mexico City, Mexico

Very Large Array Radio Telescope, New Mexico, USA

Vienna Technical Museum, Vienna, Austria

Visvesvaraya Industrial and Technological Museum, Bangalore, India

Websites

appinventor.mit.edu

digitalTrends.com

engineering.com

gatesnotes.com (blog of Bill Gates)

gizmodo.com

inventNET.com (Inventors Network)

inventorsmuseum.com

iphoneincanada.ca/apple/steve-jobs-archive

mercedes-benz-publicarchive.com (Digital Archives of Mercedes-Benz Classic)

museumoffailure.com

museumofinterestingthings.org

tech2.com

techCrunch.com

theNextWeb.com

theVerge.com

unesco.org/new/en/nikola-teslas-archive

w3.org

web.mit.edu/invent (MIT Invention Dimension)

wired.com

INDEX

Cataloging-in-Publication Data has been applied for and may be obtained from the Library of Congress.

ISBN: 978-1-62795-165-4

Series Concept and Direction: Jeanette Limondjian
Design: Bradbury and Williams
Editor: Meredith MacArdle
Proofreader: Julia Adams
Picture Research: Clare Newman
Consultant: Giles Sparrow
Cover Design: Igor Satanovsky

Publishers' Note: While every effort has been made to ensure that the information herein is complete and accurate, the publisher and authors make no representations or warranties either expressed or implied of any kind with respect to this book to the reader. Neither the authors nor the publisher shall be liable or responsible for any damage, loss, or expense of any kind arising out of information contained in this book. The thoughts or opinions expressed in this book represent the personal views of the authors and not necessarily those of the publisher. Further, the publisher takes no responsibility for third party websites or their content.

SHELTER HARBOR PRESS
603 West 115th Street Suite 163
New York, New York 10025

For further information please contact:
info@shelterharborpress.com

Printed and bound in China.

10 9 8 7 6 5 4 3 2 1

PICTURE CREDITS

MAIN BOOK
Alamy: Auscape International Pty Ltd 10cr, Chronicle 37b, 73c, Classic Image 13, DPA Picture Alliance Archive 59tl, Mark Dunn 49b, Everett Collection Historical 137bl, Everett Collection Inc 64t, 83tr, FLHC37 58b, Granger Historical Picture Archive 46b, Idealink Photography 139tl, Interfoto 64/65, ITAR-TASS News Agency 81t, 115t, John D. Ivanko 98t, Robert Kawka 22b, Lucky Step 48 116/117, Mopic 126t, Jeff Morgan 05 68tl, Trinity Mirror 89b, Nature 125t, North Wind Picture Archives 48b, Pictorial Press Ltd 68tr, 72bl, Science History Images 22t, 33cr, 43t, 49t, 76b, Space X 81cr, The Print Collector 52br, Universal Images Group North America LLC 51tl, UPI 89cl, Jim West 5, Jim West 112b, World History Archive 2, 32t, 50/51, 50bl, 52tr; **Getty Images:** Hulton Archive 84b, Popperfoto 73br, SSPL 3bl, 36t, 82br, Dong Wenjie 113b; **Library of Congress:** 58t; **MEPL:** 6c, 29t; **Science Photo Library:** 107cr, Ann Ronan Picture Library 53t, Nicolle R. Fuller 131b, Mikkel Juul Jensen 66t, KTS Design 9br, Sputnik 130tc; **Shutterstock:** Aapsky 8tl, Mark Agnor 104b, All is Magic 117br, Alliure Productions 75br, Popel Arseniy 130cr, Barbara Ash 4tr, 70tl, Asharkyu 96b, Alexander Babich 8cl, Jim Barber 39t, Gabriela Beres 25b, Bessarab 129t, Tyler Boyes 95t, Catmando 100b, CI Photos 129c, R. Classen 9cl, Colacat 21b, Jesus Cobaleda 10b, Cobalt 88 9cr, Keri Drovi 93b, Esteban De Armas 12tr, Distintive Shotsq 70bl, Dja65 38r, Adrian Domundo 85b, Dotted Yeti 131t, e71lane 127t, MDV Edwards 117t, eFesenko 6/7, 53b, EmBaSy 8tr, Finlandi 62b, Freestyle Images 128t, Garanga 31t, Gilmanshin 19b, Gorosan 60tr, Nikoletta Lia Huhari 90tr, Tim Jenner 60bl, Yien Jeon 114, Kaving 87cl, Kletr 19t, Gorod en Koff 127cr, Ekaterina Kondratova 130bl, ktsdesign 127b, Peter Lorimer 27t, Frederic Legrand-COMEO 109l, 138bl, Terence Mandoza 70cr, Jirapong Manustrong 113t, Said Marroun 103b, Marzolino 23b, 29b, 57cr, Matriushka 24b, Luzimag Andres Membrive 34t, Memory Past bkk 68bl, Metamorworks 112t, Tony Moran 106/107b, Morphart Creation 28b, 35t, Nikkytok 87br, Nataliia Natykach 8b, Nuntiya 21t, Writ Olszewski 90bl, Lefteris Papaulakis 55cr, Andrey Popov 105, Andrew Roiland 24cl, J. Roberts Williams 92b, Fouad A. Saad 82cl, Dario Sabljak 129br, Sammy33 31b, Sashkin 8cr, Serg Bob 111bl, Shade Design 120b, Solar Seven 103t, Solar Studio 94r, Space Wind 128b, Stanslavs 102t, J Stone 138br, Studio Touch 57tl, Stunning Art 97t, Sunshine Art 111cr, Andrey Suslov 9bl, Travel Mania 86t, Triff 39b, Twinlynx 14r, V41D Afsahi 33t, Ververidis Vasilis 102b, Venus Angel 18, Verona Studio 95b, Cat Walker 139tr, WIP-Studio 110t, Ymgerman 101tr, Zigres 14l; **Stocksy:** Colin Anderson 3;

The Wellcome Library, London: CC BY 4.0 38l, 46t, 54t, 67cr, 76t, 132bl, 132br, 134tl, 134tr, 135tl, 136bl; **Wikimedia Commons:** 79b, 104t, CC BY 2.0 101b, CC BY SA 3.0 17b, CC BY 1.0 92t, CC BY 2.0 85cr, 91bl, 99t, 108t, 111t, 139br, CC BY SA 2.0 45b, 59br, 28trt, 98b, CC BY 2.5 15b, CC BY SA 2.5 11cl, 52bl, CC BY 3.0 11br, 25t, 84t, 115b, 125b, CC BY 4.0 121b, 126b, 134bl, CC BY SA 3.0 10cl, 16l, 23t, 40b, 44t, 45t, 56b, 65tl, 67b, 71tr, 72br, 78b, 79t, 86br, 88t, 108b, 120t, 125cr, 137br, CC BY SA 4.0 32b, 54bl, 61b, 64b, 99b, 116, 124t, 138tl, 138tr, CC0 26-71b, 96t, CC0 1.0 15cl, 27b, CC0 1.0 12bl, GNU Free Documentation License 83cl, NASA 124b, Public Domain incl, U.S. 3, 7br, 7tl, 11tl, 15cr, 16r, 17t, 20, 26l, 33bl, 34b, 36b, 37t, 40cr, 40tl, 41br, 41cl, 41tr, 42, 43b, 44l, 47b, 47t, 40t, 51br, 51tr, 55bl, 55t, 56t, 57br, 61c, 61t, 62t , 63, 63t, 65cr, 66b, 69b, 69t, 72/73, 74, 74b, 75bl, 75t, 77b, 77t, 78t, 79cr, 80/81, 80l, 83br, 85t, 87t, 89tr, 91tr, 93t, 94l, 97b, 100t, 106, 106-107t, 109r, 1110b, 132tr, 133bl, 133br, 133tl, 133tr, 134br, 135bl, 135br, 135tr, 136bc, 136tl, 136tr, 137tl, 137tr.

TIMELINE:
Getty Images: SSPL; **NASA:** Public Domain incl, U.S.; **Shutterstock:** Altrendo Images, Andrey VP, dani3315, Juan Ci, EcoPrint, Yien Jean, Morphart Collection, Stunning Art; **The Wellcome Collection, London:** CC BY 4.0; **Wikimedia Commons:** CC BY 2.0, CC BY SA 2.0, CC BY SA 2.5, CC BY SA 3.0, CC BY SA 4.0, Public Domain incl, U.S.

REVERSE TIMELINE:
Alamy: Chronicle, Interfoto, Haiyin Wang, World History Archive; Library of Congress; **NASA:** GCTC/Andrey Shelepin; **Reaction Engines Ltd**; **Science Photo Library:** David A. Hardy; **FUTURES:** 50 years in space; **Wikimedia Commons:** CC BY 4.0, Public Domain.

Author biographies

Tim Cooke
Tim Cooke has written more than 250 books on a wide range of subjects. He is particularly interested in history, and the way that science and invention have changed our lives over time.
(Articles 31, 45, 59, 60, 64, 77, 78, 82, 84, 100)

Tim Harris
In more than 20 years, Tim Harris has written dozens of books about science and nature for children and young people, and has a particular interest in the history of scientific ideas. He has a fascination with making complex concepts digestible for enquiring minds.
(Articles 1–25)

Tom Jackson
Tom Jackson specializes in recasting science and technology into lively historical narratives. After more than 20 years of writing, Tom has uncovered a wealth of stories that help create new ways to enjoy learning about all kinds of subjects.
(Articles 26, 29, 40, 42, 47, 54, 55, 58, 66, 69, 71, 73, 74, 79, 81, 83, 86–89, 91, 94, 96, 97)

Robert Snedden
Robert Snedden has been involved in publishing for over 40 years, researching and writing science and technology books for young people on topics ranging from ancient technology to space exploration, engineering, computers, and the Internet. He has also written and contributed to books for an adult audience on mathematics, engineering, technology, physics, biology, and evolution.
(Articles 27, 28, 30, 32–39, 41, 43, 44, 46, 48–53, 56, 57, 61–63, 65, 67, 68, 70, 72, 75, 76, 80, 85, 90, 92, 93, 95, 98, 99)

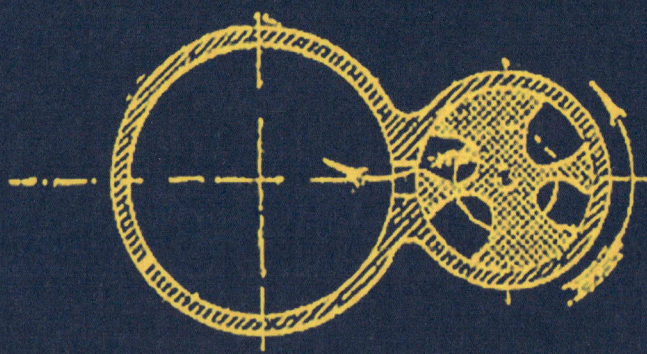

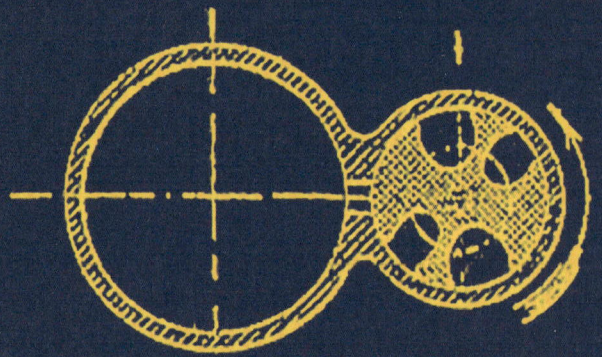